Fundamentals of the Pure Spinor Formalism

This work has been accomplished at the Institute for Theoretical Physics (ITFA) of the University of Amsterdam (UvA) and is financially supported by a Spinoza grant of the Netherlands Organisation for Scientific Research (NWO).

Lay out: Joost Hoogeveen, typeset using LaTeX
Cover design: René Staelenberg, Amsterdam
Cover illustration: Wayne Ruston

ISBN 978 90 5629 641 4
NUR 910

Fundamentals of the Pure Spinor Formalism

Academisch Proefschrift

ter verkrijging van de graad van doctor

aan de Universiteit van Amsterdam

op gezag van de Rector Magnificus

prof. dr. D. C. van den Boom

ten overstaan van een door het college voor promoties ingestelde

commissie, in het openbaar te verdedigen in de

Agnietenkapel

op vrijdag 9 juli 2010, te 12:00 uur

door

Joost Hoogeveen

geboren te Leidschendam

Promotiecommissie

Promotor

prof. dr. R. H. Dijkgraaf

Co-Promotor

dr. K. Skenderis

Overige leden

prof. dr. N. J. Berkovits

prof. dr. J. de Boer

prof. dr. P. van Nieuwenhuizen

dr. M. M. Taylor

prof. dr. E. P. Verlinde

Faculteit der Natuurwetenschappen, Wiskunde en Informatica

Publications

This thesis is based on the following publications:

1. Joost Hoogeveen and Kostas Skenderis,
   BRST quantization of the pure spinor superstring,
   JHEP **0711** (2007) 081 [arXiv:0710.2598].
2. Joost Hoogeveen and Kostas Skenderis,
   Decoupling of unphysical states in the minimal pure spinor formalism I,
   JHEP **1001** (2010) 041 [arXiv:0906.3368].
3. Nathan Berkovits, Joost Hoogeveen, Kostas Skenderis,
   Decoupling of unphysical states in the minimal pure spinor formalism II,
   JHEP **0909** (2009) 035 [arXiv:0906.3371].

Contents

Preface

This thesis is the tangible result of my years as a PhD student at the Institute for Theoretical Physics of the University of Amsterdam. During this period I have, in cooperation with my co-authors, published three papers on the pure spinor formalism, which have already been mentioned in this preliminary part. While the content of these papers forms an important part of this work, the latter also contains an introduction into all the ingredients that are necessary to fully appreciate the results of the papers. In writing this introduction I only have assumed that the reader is familiar with quantum field theory and general relativity, so in particular knowledge of string theory is not absolutely necessary, although readers who have read some textbook material on perturbative string theory will find this thesis easier to read.

Argueably the best one line description of the first chapter is: "the shortest path from quantum field theory and general relativity to the pure spinor formalism". In this chapter perturbative string theory is introduced along with its motivations. I have done this in a way that puts emphasis on the parts that are relevant for the pure spinor formalism, which is the most recent string theory formalism. This formalism is introduced in chapter two, where I will demonstrate that the pure spinor formalism distinguishes itself from the two other string theory formalisms (RNS and Green-Schwarz) by the fact that important symmetries of the theory (Lorentz invariance and supersymmetry) are manifest. This has a simplifying effect on amplitude computations and indeed the pure spinor formalism has proved to be more powerful than the other two formalisms. The next chapter contains details of the arguments used in chapter two and sets the stage for the derivations in the next two chapters.

The last couple of chapters before the conclusion is based on my three papers, which all deal with fundamental issues of the pure spinor formalism. The first provided a first principles derivation for the amplitude prescription of the pure spinor formalism. The other two papers contain the proof of an important property that any quantum theory must have, which involves unphysical states. These are states that one includes in the theory at an intermediary stage in order to preserve important symmetries, but should not be part of the physical spectrum of the theory. Thus in the end the theory should be such that if one scatters a number of physical states, unphysical states will not be produced, in other words unphysical states must decouple. The proof of this decoupling in the case of the pure spinor formalism is the subject of the fifth chapter. More precisely this chapter contains the proof for one of the two versions of the pure spinor formalism, the so-called minimal formalism. Decoupling of unphysical states in the other version is trivial to prove and will therefore not be discussed at length.

Joost Hoogeveen
Amsterdam, Netherlands
June 2010

Chapter 1

String theory

1.1 Motivations for supersymmetric string theory

High energy physics describes the most fundamental processes that occur in nature. These comprise the interaction of elementary particles, which include for instance electrons, protons and neutrinos. The theoretical description of these processes began in the beginning of the twentieth century when, among others, Bohr, Heisenberg and Schrödinger wrote down the laws and principles of quantum mechanics. These could explain the surprising outcome of the double slit experiment and the peculiar nature of the hydrogen spectrum. In the succeeding decades much more was learnt about elementary particle physics, in particular how to combine quantum mechanics with the theory of relativity as developed by Einstein in the early 1900's. The replacement of all individual particles by a much smaller number of fields, one for each kind of particle, lied at the heart of the progress in relativistic quantum mechanics, which goes under the name of quantum field theory. The introduction of fields allows for the creation and annihilation of particles that any relativistic theory necessarily contains, since kinetic energy is exchangeable with mass, as expressed by Einstein's famous formula $E = mc^2$. Around 1970 all the work on quantum field theory culminated in a realistic model that describes interactions of elementary particles with a stunning accuracy, the standard model[1]. This model provides theoretical explanations and predictions for the behaviour of elementary particles under the influence of three of the four (known) forces of nature: electromagnetism, the weak (nuclear) force and the strong (nuclear) force. It accounts for all observed particles, which is non trivial from a theoretic viewpoint because if the number of fields of some kind is different from a certain number, the theory would not be unitary, i.e. it predicts

[1] An excellent reference for quantum field theory is [1]. This book contains the details of the field theory arguments used in this section.

negative probabilities. It also predicts the existence of one particle that has not been unobserved (yet), the Higgs boson. This fact and other experimental data put strong constraints on the allowed values of its mass. If the Higgs boson exists in nature its mass must lie in the range

$$115 \text{ GeV}c^{-2} < m_H < 500 \text{ GeV}c^{-2}. \tag{1.1}$$

The failure to detect the Higgs boson in particle accelerators to date gives the lower bound. A Higgs mass higher than the upper bound would have resulted in non standard model physics at measurable scales. If one assumes the standard model describes nature all the way up to the Planck scale,

$$M_P = \left(\frac{G_N}{\hbar c}\right)^{-\frac{1}{2}} \sim 10^{19} \text{ GeV}c^{-2}, \tag{1.2}$$

the upper bound on the Higgs mass drops to approximately 180 GeVc^{-2}. As will become apparent in the next paragraph it is interesting to note how the physical value of the Higgs mass is obtained in the standard model. One starts with a bare mass, m_H^0, and then adds quantum corrections to it:

$$m_H^2 = (m_H^0)^2 + \delta m_H^2. \tag{1.3}$$

The quantum correction, δm_H^2, can be expanded in powers of the cut off of the theory. This is the scale up to which one assumes the theory is valid.

$$\delta m_H^2 = c_1 \Lambda^2 + c_2 m_H^2 \log \frac{\Lambda}{m_H} + \cdots . \tag{1.4}$$

The first term in this expansion comes completely from the diagram in figure 1.1.

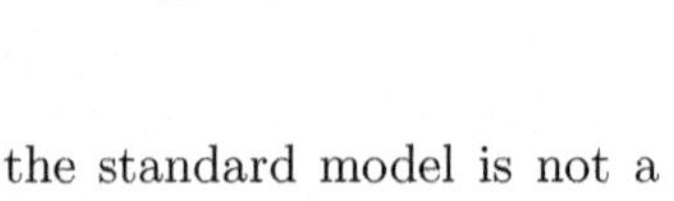

Figure 1.1: *Higgs mass renormalisation diagram*

In spite of its success the standard model is not a complete theory, the most serious problems are listed below.

- The standard model does not include *gravity*. This is not a problem for describing earth based particle accelerator experiments like the LHC because gravity is very weak compared to the other three forces. The lack of a theory of quantum gravity does, however, make for instance the description of the universe just after (and perhaps before) the big bang impossible. Other motivations to develop such a theory include the desire for a microscopic description of black holes and the search for dark, i.e. non standard model, matter candidates. The existence of dark matter can be inferred from cosmological observations on the expansion of the universe. These observations give an estimate for the total mass in the universe. The standard model particles can only account for four percent of the total mass. Hence there is much more to our universe than the standard model.

- The standard model is one element of an *18 parameter family* of theories. The value of these parameters must be obtained from experiment. There is no physical principle to determine them.

- When the underlying mass scale of a theory is of a completely different order than the masses measured in experiments a theory is said to have a *hierarchy problem*. This is considered unnatural by many, because the quantum corrections to the bare mass values are of the order of the fundamental parameters (cf. (1.4)), which implies that the bare mass m_H^0 has to be "fine-tuned" with the utmost precision to give the observed physical mass. The standard model is an example of a such a theory. There are two choices for the fundamental mass scale. The first one is the Planck mass, which is the only mass that can be constructed out of constants of nature. Furthermore it is the scale at which gravity starts to play a role. The second choice is the grand unification mass, which is the scale where the three coupling constants of the standard model (roughly) meet:
$$M_{GUT} \sim 10^{16} \text{ GeV}c^{-2}. \tag{1.5}$$
Clearly both these scales are of completely different order than the Higgs mass, m_H.

  The hierarchy problem can also be posed as the lack of an explanation for the fact that gravity is so weak compared to the other three forces. More explicitly why do standard model particles carry electric charge of order one and gravitational charges of order 10^{-19}, in dimensionless units.

These three problems have various resolutions, a number of which will be discussed in this chapter.

The introduction of a new symmetry to the theory, *supersymmetry*, is a partial resolution to the hierarchy problem. This is a symmetry that relates bosons to fermions, in particular every standard model particle has a superpartner in a supersymmetric theory. To see the effect on δm_H^2 first note that a supersymmetric theory has much more (Lorentz) scalars because all the fermions have superpartners. One can show that the Λ^2 term coming from the diagram in figure 1.1, with the Higgs in the loop replaced by an arbitrary scalar cancels exactly against a Λ^2 term coming from the diagram in figure 1.2, where the fermion running in the loop is the superpartner of the boson that replaces H in the loop in figure 1.1. The leading order

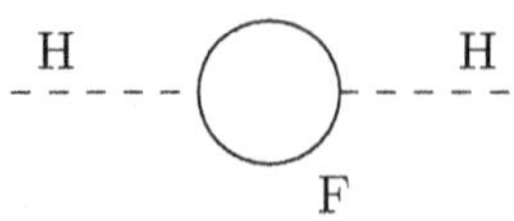

Figure 1.2: *Higgs mass renormalisation diagram, F = fermion*

correction goes like $m_t \log \frac{\Lambda}{m_t}$, where m_t is the mass of the heaviest particle. This solves the naturalness part of the hierarchy problem, there is no need for fine tuning anymore. The huge gap between the Planck scale and the mass scale of elementary particles is, however, still unexplained. Nonetheless supersymmetry can be used as a guidance principle in the search for a more complete theory of nature.

The most obvious (and naive) way to construct a theory of all four forces follows the same recipe that was used to build the standard model. This means start with the classical Lagrangian and quantise it. The action that describes gravity at the classical level is the Einstein-Hilbert action:

$$S_{EH} = \frac{c^3}{16\pi G_N} \int d^4x \sqrt{-g} R, \tag{1.6}$$

where $g_{\mu\nu}$ is the spacetime metric, R is its Ricci scalar and

$$g \equiv \det(g_{\mu\nu}). \tag{1.7}$$

Quantising the above action (plus terms that couple gravity to standard model fields) leads to infinities that cannot be handled in a sensible way, in other words the Einstein-Hilbert action is *non renormalisable.* An easy check for non renormalisability of an interaction term is provided by calculating the mass dimension of the coupling constant. If this dimension is negative the interaction term is non renormalisable. As an example consider ϕ^4 theory:

$$\int d^4x \left(\partial_\mu \phi \partial^\mu \phi + m^2 \phi^2 + \lambda \phi^4\right). \tag{1.8}$$

The mass dimension of ϕ can be inferred from either of the first two terms:

$$[\phi] = \frac{1}{2}(-[d^4x] - [m^2]) = \frac{1}{2}(4-2) = 1. \tag{1.9}$$

This leads to $[\lambda] = 0$, hence ϕ^4 theory is (at least superficially) renormalisable. To determine the mass dimension of G_N note that $[g_{\mu\nu}] = 0$ as follows from

$$ds^2 = g_{\mu\nu} dx^\mu dx^\nu. \tag{1.10}$$

Since the Ricci scalar contains two spacetime derivatives and a number of $g_{\mu\nu}$'s, its mass dimension is 2. This implies

$$[G_N] = -2. \tag{1.11}$$

This is the dimension a coupling constant in a ϕ^6 term would have, which we know leads to a non renormalisable theory. An example of a divergent Feynman diagram that cannot be dealt with in case of a coupling constant with negative mass dimension is depicted in figure 1.3. The Feyman rules dictate to integrate over the position

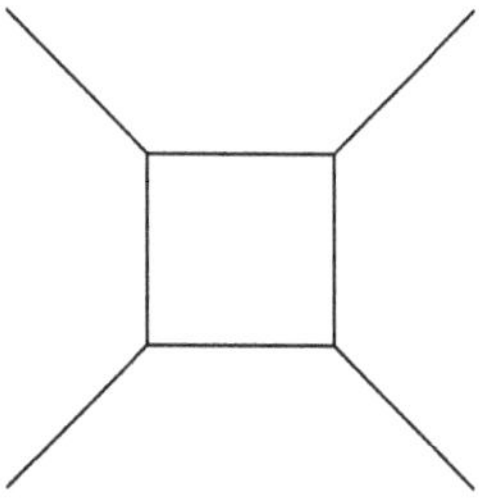

Figure 1.3: *Graviton loop*

of the vertices of the above diagram. The divergence comes from the region when the four vertices are very close to each other. Within the confines of quantum field theory it is very hard to remove this divergence in a Lorentz invariant way. One of the basic ideas of string theory is to smear out this interaction over a larger region.

The mere fact that an interaction term is non renormalisable does not rule it out as part of a realistic model. It could be that the divergence is an artifact of perturbation theory and that all physical quantities are finite when the exact theory is considered. Technically speaking such a theory would have a *non trivial UV fixed point.* Note, however, that finding such a point would not solve all the problems of the standard model mentioned above.

Before introducing string theory there is one more ingredient of quantum field theory that needs to be mentioned. It is a very elegant and powerful method, developed by Feynman, to describe and predict the scattering of elementary particles. In the case of an elementary particle he proposed that if its position at some time t_0 is given by x_0, it could evolve to any other position x_1 at time t_1. The probability of this process can be calculated by assigning a weight to all paths with the given initial and final data, even very non-classical ones. As an example of such a path one can think of a path that goes to the sun and back, with the initial and final position within the same room. The probability of the particle evolving from x_0 to x_1 is given by the sum, or integral, of the weights of all possible paths. The very non-classical paths have very small weights so that they hardly contribute. This is a key concept in the standard model and quantum field theory in general. Its great importance to theoretical physics is also demonstrated by the pivotal role that *Feynman's path integral* plays in string theory, a point that will be discussed in detail in due course.

In string theory elementary particles are no longer thought of as point particles but as *strings*. More explicitly as one dimensional objects that can either be open or closed. Different excitations of the string correspond to different particles. To describe the evolution of strings the path integral method is used:

$$P(x_i(t_i,\sigma) \to x_f(t_f,\sigma)) = \int \mathcal{D}p W(p), \tag{1.12}$$

where P is for probability and σ is a variable that parametrises the string. The integration variable p denotes an arbitrary path, which is a two dimensional surface in spacetime, the worldsheet. This worldsheet is parametrised by two worldsheet coordinates σ^1 and σ^2. Its embedding in spacetime is given by $X^\mu(\sigma^1,\sigma^2)$. W is the weight that needs to be specified. Feynman specified this weight as $e^{iS/\hbar}$ for some action S, so that all weights have the same magnitude and only differ by their phases. Contributions of highly non-classical paths are suppressed by interference. To compute the probability of two initial states to interact and produce two final states, Feynman's path integral principle tells you to sum over all possible paths that connect the four states. This path integral splits up in different terms, the diagrams. Figure 1.4 shows this process in string theory. Note that each diagram

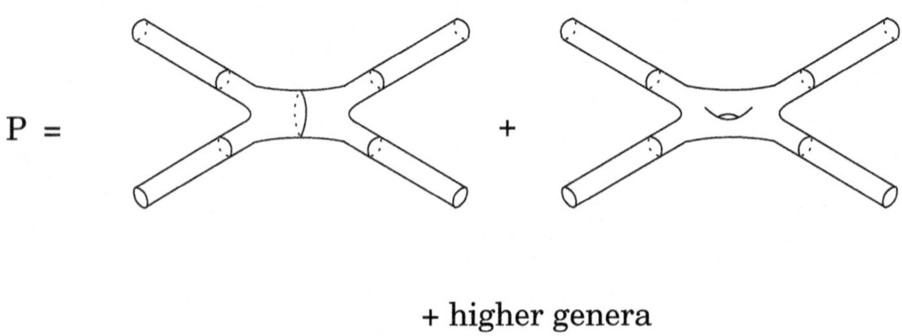

Figure 1.4: *Scattering in closed string theory*

depicts a sum over all embeddings with a particular genus (the number of handles). It is not possible to call one point on the worldsheet in figure 1.4 the interaction point. This is what is meant by smearing out the interaction. Also note this is an interaction in closed string theory, which will be described first. Thereafter open strings will be discussed briefly.

1.2 Bosonic string theory

The easiest choice one can make for the weight is the exponential of the area of the worldsheet of the string. In general the area of a surface is calculated by

$$A = \int_{\Sigma_g} d^2\sigma \sqrt{-\mathrm{det} h_{ab}}, \tag{1.13}$$

where h_{ab} is the metric on the surface, denoted by Σ_g where g is the genus. The torus has genus one for instance. If the surface is embedded in space time this h is the induced metric and the weight W becomes

$$W(X(\sigma)) = e^{\frac{i}{\hbar} S_{NG}(X)}, \tag{1.14}$$

where S_{NG} is the Nambu Goto action[2] which reads

$$S_{NG}(X) = \frac{1}{2\pi\alpha'} \int d^2\sigma (-\mathrm{det}[\partial_a X^\mu \partial_b X_\mu])^{\frac{1}{2}}. \tag{1.15}$$

The constant α' has the dimensions of a length squared and it is interpreted as the square of the string length. The square root in the Nambu Goto action makes computation based on it cumbersome. Such a situation is often encountered in theoretical physics. The way forward is introducing extra degrees of freedom and (gauge) symmetries at the same time. In this case this results in the Polyakov action:

$$S_P = \frac{1}{2\pi\alpha'} \int d^2\sigma \sqrt{g} g^{ab} \eta_{\mu\nu} \partial_a X^\mu \partial_b X^\nu, \tag{1.16}$$

The field g_{ab} constitutes three extra degrees of freedom and it has the interpretation of worldsheet metric. The first gauge symmetry is two dimensional diffeomorphism invariance:

$$\sigma^a \to (\sigma')^a(\sigma), \tag{1.17}$$

which induces transformations on g and X:

$$\delta X^\mu = v^a \partial_a X^\mu, \quad \delta g_{ab} = -\nabla_{(a} v_{b)}, \tag{1.18}$$

where $v(\sigma)$ is a parameter of infinitesimal diffeomorphisms. The second symmetry is Weyl invariance:

$$g_{ab} \to e^{2\omega(\sigma)} g_{ab}. \tag{1.19}$$

Together these are three gauge invariances, equal to the number of introduced degrees of freedom. Furthermore one can show the two actions are classically equivalent by examining the field equations.

[2]From this point onwards natural units will be used, i.e. $c = \hbar = 1$

The path integral based on the Polyakov action is given by

$$P = \int \mathcal{D}X\mathcal{D}g e^{iS_P}. \tag{1.20}$$

One can show by performing a Wick rotation (cf. [2]) that the above path integral is equal to

$$\int \mathcal{D}X\mathcal{D}g e^{-S_P}, \tag{1.21}$$

where the integration is over Euclidean worldsheet metrics g and over all embeddings X in target space. In this form without the i in the exponent the functional integrations tend to be better behaved.

The integrations in (1.20) are over function spaces, the space of all embeddings $X^\mu(\sigma)$ and all worldsheet metrics $g_{ab}(\sigma)$. Since these are not ordinary integrals it needs to be specified how these integrals are evaluated. As an example consider

$$\int \mathcal{D}X e^{-S_P} = \int \mathcal{D}X e^{\frac{1}{2\pi\alpha'}\int d^2\sigma\sqrt{g}X^\mu\nabla^2X_\mu}, \tag{1.22}$$

where compared to (1.20) the integration over the worldsheet metric is omitted. The space of all functions X^μ is spanned by the eigenstates of the operator ∇^2, which are denoted by $X_I(\sigma)$:

$$X^\mu(\sigma) = \sum_I x_I^\mu X_I(\sigma). \tag{1.23}$$

The eigenstates satisfy

$$\nabla^2 X_I = -\lambda_I^2 X_I, \quad \int d^2\sigma\sqrt{g}X_I X_J = \delta_{IJ}. \tag{1.24}$$

Integrating over all possible fields $X^\mu(\sigma)$ is the same as integrating over all possible coefficients x_I^μ. Therefore the functional measure $\mathcal{D}X$ can be replaced by an infinite product of ordinary integration measures:

$$\mathcal{D}X = \prod_{I,\mu} dx_I^\mu. \tag{1.25}$$

The integral in (1.22) can now be evaluated as

$$\int \mathcal{D}X e^{-S_P} = \int \prod_{I,\mu} dx_I^\mu e^{-\frac{1}{2\pi\alpha'}\sum_{I\neq 0}\lambda_I^2 x_I^2} = \prod_{I\neq 0}\left(\frac{2\pi^2\alpha'}{\lambda_I^2}\right)^{\frac{d}{2}} \int dx_0^\mu =$$

$$\det{}'\left(\frac{-\nabla^2}{2\pi^2\alpha'}\right)^{-\frac{d}{2}} \int dx_0^\mu, \tag{1.26}$$

where the prime on the determinant denotes the omission of the zero mode. This mode is annihilated by ∇^2 and therefore there is no Gaussian for x_0^μ. The notion of a functional determinant, such as the determinant in (1.26), can be made more precise [3], however it often suffices to include them in the overall factor. The integration over the zero modes gives the volume of spacetime in this case. However in a typical string theory computations there is an insertion $e^{ik\cdot x_0}$, so that the zeromode integral gives $\delta^d(k)$. It will become clear in due course that k denotes the sum of the momenta of the external particles.

Quantisation of gauge invariant actions

The path integral in theories with gauge invariance typically diverges badly. This is easily seen by splitting the measure of the path integral up in variables that parametrise gauge transformations and those that parametrise the physical components of the fields:

$$Z = \int \mathcal{D}\phi e^{-S} = \int \mathcal{D}\phi_{gauge}\mathcal{D}\phi_{phys}e^{-S} = \left(\int \mathcal{D}\phi_{gauge}\right)\int \mathcal{D}\phi_{phys}e^{-S} = \infty. \tag{1.27}$$

The proper definition of the path integral in gauge theories is

$$Z = \int \frac{1}{Vol\ G}\mathcal{D}\phi e^{-S[\phi]}. \tag{1.28}$$

This can be made more precise by using a method developed by *Faddeev and Popov.* The first step is writing 1 in a very special way:

$$1 = \Delta_{FP}(\phi)\int \mathcal{D}\zeta\mathcal{D}\tau\delta(\phi^\zeta - \hat{\phi}(\tau)), \tag{1.29}$$

where ζ parametrises a single gauge orbit and τ parametrises the space of gauge orbits, as depicted in figure 1.5. The integral in (1.29) is nonvanishing because there is always a choice for ζ and τ such that $\phi^\zeta = \hat{\phi}(\tau)$. Since the integral in (1.29) is non-vanishing it makes sense to regard (1.29) as the definition of $\Delta_{FP}(\phi)$. It will be explicitly computed in a number of cases below. Inserting the 1 into Z gives

$$Z = \int \frac{1}{Vol\ G}\mathcal{D}\phi\mathcal{D}\zeta\mathcal{D}\tau\Delta_{FP}(\phi)\delta(\phi^\zeta - \hat{\phi}(\tau))e^{-S[\phi]} = \tag{1.30}$$

$$\int \frac{1}{Vol\ G}\mathcal{D}\zeta\mathcal{D}\phi^\zeta\mathcal{D}\tau\Delta_{FP}(\phi)\delta(\phi^\zeta - \hat{\phi}(\tau))e^{-S[\phi^\zeta]}.$$

In the second line the measure for the classical fields $\mathcal{D}\phi$ has been replaced by $\mathcal{D}\phi^\zeta$. This gauge invariance of the measure often holds, but should in principle be checked when one is applying the BRST quantisation procedure. Gauge invariance of the classical action was used to replace ϕ by ϕ^ζ in the action. Furthermore the Faddeev Popov determinant is gauge invariant. In order to see this one has to show the value of the integral in (1.29) does not change when ϕ is replaced by a gauge transformed version ϕ^{ζ_1}:

$$\left(\Delta_{FP}(\phi^{\zeta_1})\right)^{-1} = \int \mathcal{D}\zeta\mathcal{D}\tau\delta((\phi^{\zeta_1})^{\zeta} - \hat{\phi}(\tau)) = \tag{1.31}$$

$$\int \mathcal{D}\zeta_2\mathcal{D}\tau\delta(\phi^{\zeta_2} - \hat{\phi}(\tau)) = \left(\Delta_{FP}(\phi)\right)^{-1}$$

where $\zeta_2 = \zeta \circ \zeta_1$ and in the second equality one has to use that integrating over all ζ_2 is the same as integrating over all ζ. Perhaps an easier way to understand the gauge invariance of the Faddeev Popov determinant is from figure 1.5. The determinant only depends on the behaviour of the field at the intersection point of the gauge orbit and the gauge slice, since the delta function vanishes away from that point. The gauge transformation will move ϕ up or down a gauge orbit, but it will

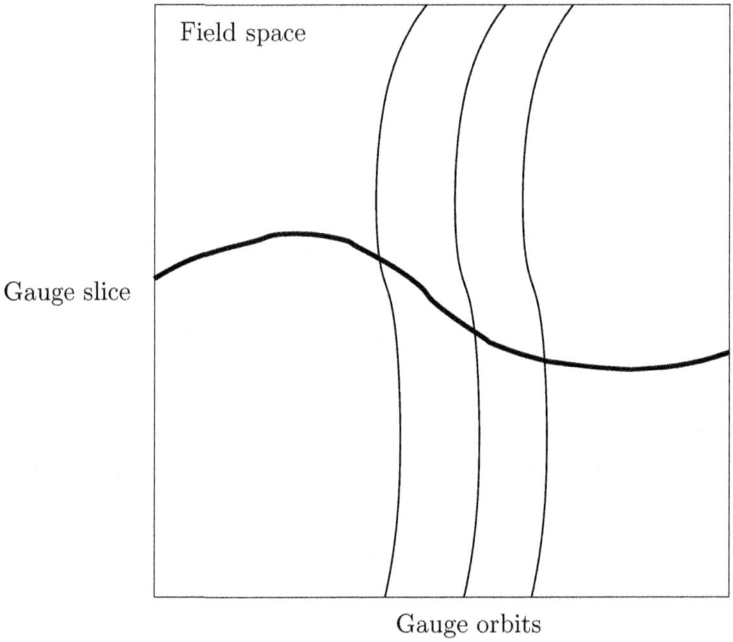

Figure 1.5: *Field space decomposes into gauge equivalent directions (vertical) and non-equivalent directions (horizontal). The delta function in* (1.29) *will only have support at the intersection of the gauge slice with the gauge orbit that contains ϕ. Therefore the Faddeev Popov determinant only depends on the gauge orbit, not on the field ϕ itself. More precisely the Faddeev Popov determinant only depends on the derivatives of ϕ at the intersection point along the gauge slice and the gauge orbit.*

not change the intersection point. Using the gauge invariance of the Faddeev Popov determinant leads to

$$Z = \int \frac{1}{Vol\ G} \mathcal{D}\zeta \mathcal{D}\phi^\zeta \mathcal{D}\tau \Delta_{FP}(\phi^\zeta)\delta(\phi^\zeta - \hat{\phi}(\tau))e^{-S[\phi^\zeta]} =$$

$$\int \mathcal{D}\phi \mathcal{D}\tau \Delta_{FP}(\phi)\delta(\phi - \hat{\phi}(\tau))e^{-S[\phi]}, \tag{1.32}$$

where ϕ^ζ was relabelled ϕ and after this modification there is no ζ dependence left hence the $\mathcal{D}\zeta$ integration cancels against the volume factor in the denominator. As a final step one can perform an integration over the classical fields to remove the delta function:

$$Z = \int \mathcal{D}\tau \Delta_{FP}(\hat{\phi}(\tau))e^{-S[\hat{\phi}(\tau)]}. \tag{1.33}$$

Both the delta function and the Faddeev Popov determinant in (1.32) can be written as functional integrals. The delta function can be rewritten as

$$\int \mathcal{D}B e^{-S_{gf}}, \quad S_{gf} = \int d^d x B_A(\phi^A - \hat{\phi}^A(\tau)), \tag{1.34}$$

where A runs over the fields that play a role in the gauge fixing. After making this restriction the number of moduli will be finite in all cases to be discussed in this thesis. The inverse of the Faddeev Popov determinant is given by

$$\Delta_{FP}^{-1}(\phi) = \int d\tau \mathcal{D}\zeta \delta(\phi^\zeta - \hat{\phi}(\tau)) = \int d\tau \mathcal{D}\zeta \delta(\phi_0 + \zeta^a \frac{\delta\phi}{\delta\zeta^a} - \phi_0 - \tau^k \frac{\partial\hat{\phi}}{\partial\tau^k}), \tag{1.35}$$

where ϕ_0 is the value of the classical fields at the intersection point of the gauge slice and the relevant gauge orbit. The truncation of the Taylor series is not an approximation. In order to see this consider the case that the classical field space is two dimensional, with one gauge direction and one modulus direction, as depicted in figure 1.5. The intersection point ϕ_0, the gauge transformed field ϕ^ζ and the gauge orbit $\hat{\phi}(\tau)$ can be written more explicitly as

$$\phi_0 = (\phi_0^1, \phi_0^2), \quad \phi^\zeta = (f(\zeta), 0) + (\phi_0^1, \phi_0^2), \quad \hat{\phi}(\tau) = (0, g(\tau)) + (\phi_0^1, \phi_0^2). \tag{1.36}$$

In this example the delta function becomes

$$\delta^{(2)}(\phi^\zeta - \hat{\phi}(\tau)) = \delta(f(\zeta))\delta(g(\tau)) = \delta(\zeta - \zeta_0)\delta(\tau - \tau_0)\frac{1}{f'(\zeta_0)}\frac{1}{g'(\tau_0)}, \tag{1.37}$$

where ζ_0 is the gauge parameter such that $\phi^{\zeta_0} = \phi_0$. This equation must have a unique solution otherwise the gauge fixing condition was erroneous. Similarly τ_0 is defined by $\hat{\phi}(\tau_0) = \phi_0$. The RHS of (1.37) can be rewritten as

$$\delta((\zeta - \zeta_0)f'(\zeta_0))\delta((\tau - \tau_0)g'(\tau_0)) = \delta^{(2)}(\zeta^a \frac{\partial\phi}{\partial\zeta^a} - \tau^k \frac{\partial\hat{\phi}}{\partial\tau^k}) \tag{1.38}$$

After this short intermezzo let us return to the Faddeev Popov determinant:

$$\Delta_{FP}^{-1}(\phi) = \int d\tau \mathcal{D}\zeta \delta(\zeta^a \frac{\delta\phi}{\delta\zeta^a} - \tau^k \frac{\partial\hat{\phi}}{\partial\tau^k}) = \int d\tau \mathcal{D}\zeta \mathcal{D}\beta e^{\beta_A(\zeta^a \frac{\delta\phi^A}{\delta\zeta^a} - \tau^k \frac{\partial\hat{\phi}^A}{\partial\tau^k})}, \tag{1.39}$$

where β_A has the same statistics as ϕ^A. Now one has to replace the fields in the path integral measure by fields of opposite statistics ($\beta_A \to b_A, \zeta^a \to c^a, \tau \to \xi$) to obtain an expression for Δ_{FP}:

$$\Delta_{FP}(\phi(\tau)) = \int \mathcal{D}b_A \mathcal{D}c^a d\xi e^{-S_{FP}}, \tag{1.40}$$

where

$$S_{FP} = \int d^d x b_A(x) \left[\frac{\delta\phi^{\zeta A}(x)}{\delta\zeta^a}(\zeta_0)c^a(x) - \frac{\partial\hat{\phi}^A(x;\tau)}{\partial\tau^\kappa}(\tau_0)\xi^\kappa(x) \right]. \tag{1.41}$$

The total amplitude is now given by

$$Z = \int \mathcal{D}\phi \mathcal{D}b \mathcal{D}c \mathcal{D}B d\tau d\xi e^{-S_P - S_{gf} - S_{FP}}, \tag{1.42}$$

where S_P is the Polyakov action, but can be any gauge invariant action. The fields b_A and c^a are the Faddeev Popov ghosts and B_A is an auxiliary field since it only appears algebraically. Furthermore b_A and B_A are tensor densities so that S_{gf} and S_{FP} are coordinate invariant. One would expect the indices A and a to run over the same values, but this need not be true as will be demonstrated below in the case of the Polyakov action.

As first realised by Becchi, Stora and Rouet (BRS) and independently by Tyutin (T) the new action is no longer invariant under the gauge transformations of S_P, but it is invariant under the BRST symmetry given by[3]

$$\delta_B\phi = \epsilon c^a \delta_a \phi, \tag{1.43}$$
$$\delta_B c^a = \epsilon f^a_{bc} c^b c^c, \tag{1.44}$$
$$\delta_B b_A = \epsilon B_A, \tag{1.45}$$
$$\delta_B B_A = 0, \tag{1.46}$$
$$\delta_B \tau^k = \epsilon \xi^k, \tag{1.47}$$
$$\delta_B \xi^k = 0, \tag{1.48}$$

where ϵ is a fermionic parameter, δ_a are the generators of gauge transformations of S_P and f^a_{bc} are the structure constants of these transformations:

$$[\delta_a, \delta_b] = f^c_{ab}\delta_c. \tag{1.49}$$

[3]When this symmetry was discovered, the gauge fixing condition did not involve moduli. Therefore the τ and ξ transformations were not part of their analysis.

By Noether's procedure one can extract a BRST charge, $\mathcal{Q}$, from the above transformations. By applying the transformations twice the reader can check this charge is nilpotent:

$$\mathcal{Q}^2 = 0. \tag{1.50}$$

With the help of the BRST operator the action in equation (1.42) can be written in a more illuminating form:

$$S = S_P + \mathcal{Q}\left[b_A(\phi^A - \hat{\phi}^A)\right], \tag{1.51}$$

where the quantity in the square brackets is called the *gauge fixing fermion*. One can see that BRST quantisation amounts to multiplying Z by a factor of the form $1 + \mathcal{Q}\Omega$, restricting the ϕ integrals to moduli space and introducing the integrations over the ghosts and auxiliary fields. The above form of Z is appealing since one can always add a $\mathcal{Q}$ exact piece to the action without changing any physical scattering processes. At a superficial level this is the idea behind BRST quantisation, one adds a $\mathcal{Q}$ exact piece to the action such that the functional integral gives a finite result.

The *spectrum* of the theory is defined as the cohomology of $\mathcal{Q}$. This means all states that are $\mathcal{Q}$ closed and states that differ by a $\mathcal{Q}$ exact state are physically equivalent:

$$\mathcal{Q}|\alpha\rangle = 0, \quad |\alpha\rangle \sim |\alpha\rangle + \mathcal{Q}|\beta\rangle. \tag{1.52}$$

One can show the first condition, together with the property (1.51), is necessary for physical amplitudes to be independent of the gauge choice and since $\langle\gamma|\alpha\rangle = \langle\gamma|\alpha\rangle + \langle\gamma|\mathcal{Q}|\beta\rangle$ for any physical state, $|\gamma\rangle$, $|\alpha\rangle$ and $|\alpha\rangle + \mathcal{Q}|\beta\rangle$ represent the same state.

The BRST transformations as given in (1.43)-(1.48) only give rise to a symmetry when the structure constants in (1.49) are really constants, i.e. they do not depend on the fields and hence they are inert under the BRST transformations. Furthermore the gauge algebra (1.49) must close, i.e. there cannot be any terms proportional to the equations of motion in the RHS of (1.49) for the the BRST transformations to be a symmetry. There exists a generalisation of BRST quantisation, developed by Batalin and Vilkovisky (BV) [4], which can also handle gauge theories with non constant "structure constants" and/or open gauge algebras, however the gauge fixing condition cannot involve a moduli space. In a follow up paper [5] the same authors have shown how to quantise gauge theories with linearly dependent gauge generators. This second paper opened up the possibility to apply BV quantisation to theories with a moduli space, like string theory. The case of the bosonic string has been worked out in [6]. The fact that (1.42) leads to a consistent quantum theory can also be derived by applying BV quantisation.

1.2.1 BRST quantisation of Polyakov action

The *gauge fixing condition* should fix all invariances. The way gauge fixing works out depends on whether the worldsheet has the topology of a sphere, a torus or a higher genus surface. On the sphere every metric is Weyl equivalent to the round metric:

$$\forall g \exists \omega(\sigma) : e^{2\omega(\sigma)} g_{ab}(\sigma) = \hat{g}_{ab}(\sigma), \tag{1.53}$$

where $\hat{g}_{ab}$ is the round metric. In coordinates that cover the entire sphere except the north pole it is given by

$$d\hat{s}^2 = e^{2\omega_0(\sigma^1,\sigma^2)}[d\sigma^{1^2} + d\sigma^{2^2}], \tag{1.54}$$

with

$$e^{2\omega_0} = \frac{4}{1+(\sigma^1)^2+(\sigma^2)^2}. \tag{1.55}$$

Furthermore, since $e^{2\omega_1} g = e^{2\omega_2} g$ if and only if $w_1 = \omega_2$, there exists a unique ω that transforms a given metric to the round metric. This fixes all the Weyl invariance. In order to fix the coordinate invariance expressing the metric in ones favourite coordinates seems to do the job. One has to check however whether there are diff $\times$ Weyl transformations that leave the metric -expressed in certain coordinates- invariant. Of course $g_{ab} d\sigma^a d\sigma^b$ is invariant under all coordinate transformations, but the question is whether there are coordinate transformations that leave $g_{ab}(\sigma^1, \sigma^2)$ invariant up to a Weyl transformation, since this would mean writing down the metric in certain coordinates does not fix all the diff $\times$ Weyl invariance. In complex coordinates,

$$z = \sigma^1 + i\sigma^2, \quad \bar{z} = \sigma^1 - i\sigma^2, \tag{1.56}$$

the metric reads

$$ds^2 = 2 g_{z\bar{z}} dz d\bar{z}, \tag{1.57}$$

where

$$g_{z\bar{z}} = \frac{1}{2} e^{\omega(z,\bar{z})}. \tag{1.58}$$

The coordinate transformations on the metric become

$$\delta g_{z\bar{z}} = (\partial_z v^z + \partial_{\bar{z}} v^{\bar{z}}) g_{z\bar{z}} = (\nabla \cdot v) g_{z\bar{z}}, \tag{1.59}$$

$$\delta g_{zz} = \partial_z v^{\bar{z}} g_{z\bar{z}}, \tag{1.60}$$

$$\delta g_{\bar{z}\bar{z}} = \partial_{\bar{z}} v^z g_{z\bar{z}}. \tag{1.61}$$

The variation of $g_{z\bar{z}}$ is a Weyl transformation for any v. The variations of g_{zz} and $g_{\bar{z}\bar{z}}$ should vanish, which implies

$$\partial_z v^{\bar{z}} = 0, \quad \partial_{\bar{z}} v^z = 0. \tag{1.62}$$

Each equation has three independent globally defined solutions on the sphere:

$$v^z = c_1 + c_2 z + c_3 z^2, \quad v^{\bar{z}} = d_1 + d_2 \bar{z} + d_3 \bar{z}^2. \tag{1.63}$$

Note that poles in z are not allowed because they give an infinity at $z = 0$. Orders above two are not allowed because they give an infinity at the north pole, $z = \infty$. These six independent vector fields represent six invariances of the metric up to Weyl (or conformal) transformations. Hence they are called conformal killing vectors (CKVs). Together they form the conformal killing group (CKG). This is the group of residual gauge invariance, after g_{ab} has been fixed. This residual invariance will be fixed in a different way, which is discussed after the paragraph on the external string states.

On the torus it is no longer true that every metric is Weyl equivalent to a given one. In this case there does exist a one (complex) parameter family of metrics such that every metric is Weyl equivalent to (exactly) one metric in this family. This family of metrics is the aforementioned moduli space of metrics on the torus. Hence the functional integral over the metric contains integration over physical modes, in contrast to the sphere. Furthermore there is residual gauge invariance on the torus. The CKG is two dimensional and the CKVs are given by the constant functions:

$$v^z = c, \quad v^{\bar{z}} = d. \tag{1.64}$$

A systematic way of analysing the CKG for surfaces of arbitrary genus is writing $0 = \delta g_{\alpha\beta}$ in a fancy way:

$$0 = \delta g_{ab} = -2(P_1 v)_{ab} + (2\delta\omega - \nabla \cdot v) g_{ab}, \tag{1.65}$$

where P_n takes symmetric traceless rank n tensors into a symmetric traceless rank $(n+1)$ tensors. They are defined by

$$(P_n(T^n))_{a_1 \cdots a_{n+1}} = \nabla_{(a_1} (T^n)_{a_2 \cdots a_{n+1})} - \frac{n}{n+1} g_{(a_1 a_2} \nabla^b (T^n)_{b a_3 \cdots a_{n+1})}. \tag{1.66}$$

By taking the trace of (1.65) one finds $2\delta\omega - \nabla \cdot v = 0$ which determines $\delta\omega$. The restriction on v is

$$(P_1 v)_{ab} = 0. \tag{1.67}$$

The number of CKVs, κ, is equal to dimension of the kernel of P_1. The actual values will be given together with the dimension of the moduli space after the next paragraph.

In order to figure out the number of moduli one looks at the number of independent metric variations that are orthogonal to all the ones in (1.65):

$$0 = (\delta' g, \delta g) \equiv \int d^2\sigma \sqrt{g} \delta' g_{ab} [-2(P_1 v)^{ab} + (2\delta\omega - \nabla \cdot v) g^{ab}]. \tag{1.68}$$

The inner product denoted by $(\cdot,\cdot)$ is defined by

$$(T, T') = \int d^2\sigma \sqrt{g} T \cdot T', \tag{1.69}$$

where T and T' are tensors of equal rank and the dot denotes contraction of all indices. Equation (1.68) is equivalent to the following two equations

$$\delta' g_{ab} g^{ab} = 0, \quad (\delta' g, P_1 v) = 0. \tag{1.70}$$

The first equation implies all variations orthogonal to diff $\times$ Weyl transformations are traceless. The second equation can be rewritten with the help of the transpose of P_n, which is defined by

$$(P_n^T(T^{n+1}))_{a_1 \cdots a_n} = \nabla^b (T^{n+1})_{ba_1 \cdots a_n}. \tag{1.71}$$

and satisfies

$$(T, P_n T') = (P_n^T T, T'). \tag{1.72}$$

The second condition on $\delta' g_{ab}$ can now be written as

$$0 = (P_1^T \delta' g, v) \; \forall v \Rightarrow P_1^T \delta' g = 0. \tag{1.73}$$

Hence the number of moduli, μ, is equal to the dimension of the kernel of P_1^T. One can prove μ vanishes on surfaces with genus two and higher [2]. The Riemann-Roch theorem for closed oriented surfaces, also proved in [2], states

$$\dim \ker P_n - \dim \ker P_n^T = (2n+1)(2-2g). \tag{1.74}$$

For the special case that $n = 1$ this gives

$$\kappa - \mu = 6 - 6g. \tag{1.75}$$

Together with the results for the torus and the sphere this leads to the table 1.1.

g	κ	μ
0	6	0
1	2	2
≥ 2	0	$6g-6$

Table 1.1: *Dimensions of the CKG and the moduli space*

The path integral for the genus g term is given by

$$Z_g = \int d^\mu \tau \frac{1}{\text{CKG}} \mathcal{D}X \mathcal{D}b \mathcal{D}c d\xi e^{\frac{-1}{2\pi\alpha'} \int_{\Sigma_g} d^2\sigma \sqrt{\hat{g}(\tau)} \hat{g}^{\alpha\beta}(\tau) \partial_\alpha X^\mu \partial_\beta X_\mu - S_{FP}[\hat{g}(\tau), b, c, \xi]}, \tag{1.76}$$

where τ parametrises moduli space. This path integral still depends on the initial and final conditions which specify the states. The BRST procedure guarantees Z is independent of the gauge fixing condition if initial and final states are physical (i.e. annihilated by the BRST operator). It is interesting to note that there is a certain $\hat{g}$ that gives rise to the genus zero diagram in figure 1.4. It is also possible to choose $\hat{g}$ to be the round metric, which simplifies computations. With this new choice the genus zero diagram is depicted in figure 1.6. The string profile of the

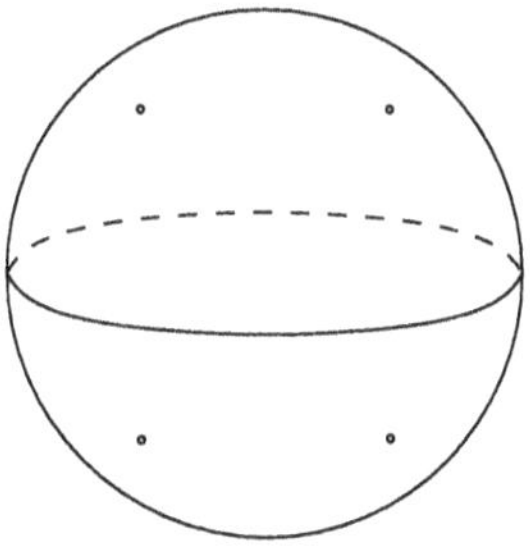

Figure 1.6: *Tree level string diagram with local vertex operators*

asymptotic states has been mapped to points. In order to understand why these are really points and not finitely sized holes consider one of the half infinite cylinders in figure 1.4. This will be mapped to a patch of the sphere that contains a puncture. The transformation on the metric that took figure 1.4 into figure 1.6 is a Weyl (or conformal) transformation on the metric. There exists a conformal transformation from the cylinder to the complex plane (the patch on the sphere) and this transformation takes the circle at the end of the cylinder to the origin of the complex plane. Hence in this gauge *string states correspond to local operators.* The explicit form of these operators will be discussed in due course.

Let us go back to arbitrary genus. It is tempting to write

$$Z_g = \int d^\mu \tau \frac{1}{\text{CKG}} \mathcal{D}X\mathcal{D}b\mathcal{D}c d\xi V_1(\sigma^1) \cdots V_N(\sigma^N) e^{-S_P[\hat{g}(\tau),X]-S_{FP}[\hat{g}(\tau),b,c,\xi]}, \quad (1.77)$$

where

$$S_P[\hat{g}(\tau), X] = \frac{1}{2\pi\alpha'} \int_{\Sigma_g} d^2\sigma \sqrt{\hat{g}(\tau)} \hat{g}^{ab}(\tau) \partial_a X^\mu \partial_b X_\mu$$

The functional integration over X is no longer constraint by boundary conditions and $V_i(\sigma^i)$ are the vertex operators and the label i represents the kind of particle and its momentum. They are inserted at arbitrary positions σ^i on the worldsheet. However, like in every gauge theory only gauge invariant objects are of physical interest and $V_i(\sigma^i)$ is not invariant under coordinate transformations. Therefore a better choice for the vertex operators is

$$\int d^2\sigma\sqrt{g}V_i(\sigma). \tag{1.78}$$

This results in

$$Z_g = \int d^\mu\tau \frac{1}{\text{CKG}}\mathcal{D}X\mathcal{D}b\mathcal{D}cd\xi d^N\sigma^i\sqrt{\hat{g}(\sigma^1)}V_1(\sigma^1)\cdots\sqrt{\hat{g}(\sigma^N)}V_N(\sigma^N)e^{-S_P-S_{FP}}. \tag{1.79}$$

Since the σ^i's appear as integration variables in the path integral they can be interpreted as quantum mechanical degrees of freedom. It is these constant fields one can use to fix the residual gauge invariance consisting of the CKG, which only plays a role on the sphere and the torus. More explicitly the gauge fixing action that fixes the entire gauge group including the CKG is given by

$$S_{gf} = \sum_{i=1}^{\kappa/2} B_i^a(\sigma_a^i - \hat{\sigma}_a^i) + \int d^2\sigma B_{ab}(g_{ab} - \hat{g}_{ab}(\tau)). \tag{1.80}$$

After integrating out the auxiliary field B the path integral becomes

$$Z_g = \int d^\mu\tau\mathcal{D}X\mathcal{D}b\mathcal{D}cd^{N-\kappa/2}\sigma^i\prod_{i=1}^{\kappa/2}\sqrt{\hat{g}(\hat{\sigma}^i)}V_i(\hat{\sigma}^i)\prod_{i=\kappa/2+1}^{N}\sqrt{\hat{g}(\sigma^i)}V_i(\sigma^i)e^{-S_P-S_{FP}}. \tag{1.81}$$

The Faddeev Popov action can be evaluated as

$$S_{FP} = \int d^dxb_A(x)\left[\frac{\partial\hat{\phi}^A}{\partial\zeta^a}c^a(x) + \frac{\partial\hat{\phi}^A}{\partial\tau^k}\xi^k(x)\right] = \tag{1.82}$$

$$(b,\hat{P}_1c) + (b,\xi^k\partial_k\hat{g}) + b_a^ic^a(\hat{\sigma}_i) + \int_{\Sigma_g} d^2\sigma b_{ab}\hat{g}^{ab}(\hat{\nabla}\cdot c - c_\omega)$$

After plugging this into Z a number of fields appears algebraically in the action: c_ω, the trace of b_{ab}, b_a^i and ξ^k. These can be integrated out:

$$Z_g = \int d^\mu\tau\mathcal{D}X\mathcal{D}b_{ab}\mathcal{D}c^ad^{N-\kappa/2}\sigma^i\prod_{i=1}^{\kappa/2}\sqrt{\hat{g}(\hat{\sigma}^i)}\epsilon_{ab}c^a(\hat{\sigma}^i)c^b(\hat{\sigma}^i)V_i(\hat{\sigma}^i)$$

$$\prod_{j=\kappa/2+1}^{N}\sqrt{\hat{g}(\sigma^j)}V_j(\sigma^j)\prod_{k=1}^{\mu}(b,\partial_k\hat{g})e^{\frac{-1}{2\pi\alpha'}\int d^2\sigma\sqrt{\hat{g}}\hat{g}^{ab}\partial_aX^\mu\partial_bX_\mu-(b,\hat{P}_1c)}, \tag{1.83}$$

where b_{ab} is traceless. In all gauges of the form $\hat{g} = e^\omega g_{round}$ and complex coordinates the action reads

$$S_{cg} = \frac{1}{2\pi\alpha'}\int d^2z\partial_zX^\mu\partial_{\bar{z}}X_\mu + b_{zz}\partial_{\bar{z}}c^z + b_{\bar{z}\bar{z}}\partial_zc^{\bar{z}}. \tag{1.84}$$

Equation (1.83) is the main formula in the bosonic string theory defined by the Polyakov action. This is where all computations start. In [6] the same formula was obtained using BV quantisation.

The BRST operator is classically nilpotent by construction. The word classically means that the transformations in (1.43) are nilpotent. In general this does not imply $\mathcal{Q}$ is also nilpotent quantum mechanically. This means $\mathcal{Q}^2$ vanishes inside (all) correlators:

$$\int \mathcal{D}\phi \mathcal{Q}^2 \mathcal{O}_1 \cdots \mathcal{O}_n e^{-S} = 0. \tag{1.85}$$

A very useful tool to examine the behaviour of operators inside a path integral is the operator product expansion (OPE). This is the statement that when two operators, say $\mathcal{O}_1$ and $\mathcal{O}_2$, inside a path integral are close to each other (no other operator is closer to either one than they are to each other) their product can be approximated to arbitrary accuracy by a sum of local operators:

$$\phi^i(z,\bar{z})\phi^j(z',\bar{z}') = \sum_k c^{ij}{}_k(z,\bar{z},z',\bar{z}')\phi^k(z',\bar{z}'). \tag{1.86}$$

As an example of this construction consider the free boson action:

$$S = \frac{1}{2\pi\alpha'}\int d^2z \partial X \bar{\partial} X. \tag{1.87}$$

In order to derive the OPE for $X(z,\bar{z})X(z',\bar{z}')$ consider the following total functional derivative

$$0 = \int \mathcal{D}X \frac{\partial}{\partial X(z,\bar{z})}\left(X(z',\bar{z}')\mathcal{O}_1(w_1,\bar{w}_1)\cdots\mathcal{O}_n(w_n,\bar{w}_n)e^{-S}\right), \tag{1.88}$$

where z and z' are close to each other and the operators $\mathcal{O}$ are inserted at points away from z and z'. By evaluating the RHS of (1.88) one obtains

$$0 = \int \mathcal{D}X(\delta^2(z-z',\bar{z}-\bar{z}') + \frac{1}{\pi\alpha'}\partial_z\partial_{\bar{z}}X(z,\bar{z})X(z',\bar{z}'))\mathcal{O}_1(w_1,\bar{w}_1)\cdots\mathcal{O}_n(w_n,\bar{w}_n)e^{-S} \tag{1.89}$$

One can conclude that inside correlators

$$\partial_z\partial_{\bar{z}}X(z,\bar{z})X(z',\bar{z}') = \pi\alpha'\delta^2(z-z',\bar{z}-\bar{z}'). \tag{1.90}$$

To be able to write this as a sum of local operators one needs to define the normal ordered product:

$$: X(z,\bar{z})X(z',\bar{z}') :\equiv X(z,\bar{z})X(z',\bar{z}') + \frac{\alpha'}{2}\ln|z-z'|^2. \tag{1.91}$$

Note this normal ordered product satisfies

$$\partial_z\partial_{\bar{z}} : X(z,\bar{z})X(z',\bar{z}') := 0. \tag{1.92}$$

Thus in normal ordered products the operators can be brought to the same point without encountering divergences. The XX OPE can now be written as

$$X(z,\bar{z})X(z',\bar{z}') = -\frac{\alpha'}{2}\ln|z-z'|^2 + :X(z,\bar{z})X(z',\bar{z}'):= \tag{1.93}$$

$$-\frac{\alpha'}{2}\ln|z-z'|^2 + :XX(z',\bar{z}'): +$$

$$\sum_{i=1}^{\infty}\frac{1}{i!}\left[(z-z')^i : X\partial^i X(z',\bar{z}') : + (\bar{z}-\bar{z}')^i : X\bar{\partial}^i X(z',\bar{z}') :\right].$$

An OPE is usually denoted as

$$X(z,\bar{z})X(z',\bar{z}') \sim \frac{\alpha'}{2}\ln|z-z'|^2, \tag{1.94}$$

where the $\sim$ means is equal to up to terms that are finite when $z \to z'$. In the sequel it will become clear that the poles in the OPE often contain enough information to do computations. The OPE for the Faddeev Popov ghosts can be obtained similarly:

$$b(z)c(w) \sim \frac{1}{z-w}, \quad b(z)b(w) \sim 0, \quad c(z)c(w) \sim 0 \tag{1.95}$$

$$\tilde{b}(z)\tilde{c}(w) \sim \frac{1}{z-w}, \quad \tilde{b}(z)\tilde{b}(w) \sim 0, \quad \tilde{c}(z)\tilde{c}(w) \sim 0, \tag{1.96}$$

where $b \equiv b_{zz}, c \equiv c^z, \tilde{b} \equiv b_{\bar{z}\bar{z}}, \tilde{c} \equiv c^{\bar{z}}$ and since the equation of motion for b is $\bar{\partial} b = 0$, $b(z,\bar{z})$ is denoted as $b(z)$. (Note this does not mean the path integral is only over holomorphic b's.)

In order to compute $\mathcal{Q}^2$ inside the path integral and the cohomology of $\mathcal{Q}$ it is useful to note the action in (1.84) is invariant under all holomorphic coordinate transformations:

$$z \to f(z). \tag{1.97}$$

These transformations are known as conformal transformations and the Noether charges associated to them are the Virasoro generators, which are given by:

$$L_n = \frac{1}{2\pi i}\int dz z^{n+1} T(z), \quad \tilde{L}_n = \frac{1}{2\pi i}\int d\bar{z}\bar{z}^{n+1}\tilde{T}(\bar{z}) \tag{1.98}$$

where $T \equiv T_{zz}$ and $\tilde{T}(\bar{z}) \equiv T_{\bar{z}\bar{z}}$ are the only two nonvanishing components of the stress energy tensor (i.e. the Noether current for translations). In a Weyl invariant theory the stress energy is traceless, therefore $T_{z\bar{z}} = T_{\bar{z}z} = 0$. Their explicit expressions are given by

$$T(z) \equiv T^m_{zz}(z) + T^g_{zz}(z), \quad \tilde{T}(\bar{z}) \equiv T^m_{\bar{z}\bar{z}}(\bar{z}) + T^g_{\bar{z}\bar{z}}(\bar{z}), \tag{1.99}$$

$$T^m_{zz} = -\frac{1}{\alpha'} : \partial X^\mu \partial X_\mu :, \quad T^m_{\bar{z}\bar{z}} = -\frac{1}{\alpha'} : \bar{\partial} X^\mu \bar{\partial} X_\mu :, \tag{1.100}$$

$$T^g_{zz} =: (\partial b)c : -2\partial(: bc :), \quad T^g_{\bar{z}\bar{z}} =: (\bar{\partial}\tilde{b})\tilde{c} : -2\bar{\partial}(: \tilde{b}\tilde{c} :). \tag{1.101}$$

All continuous symmetries imply the existence of Ward identities. For an arbitrary symmetry of the two dimensional action generated by a current $(j, \tilde{j})$ the Ward identity reads [2]

$$\oint_{\partial R} (jdz - \tilde{j}d\bar{z})\mathcal{O}(z_0, \bar{z}_0) = \frac{2\pi}{\epsilon}\delta\mathcal{O}(z_0, \bar{z}_0), \tag{1.102}$$

where ϵ is the parameter of the transformation and R is some small region in the two dimensional spacetime. In this thesis all symmetries are such that j and $\tilde{j}$ are conserved separately, i.e. j is holomorphic and $\tilde{j}$ is antiholomorphic. In these cases (1.102) becomes

$$\mathrm{Res}_{z\to z_0} j(z)\mathcal{O}(z_0, \bar{z}_0) + \overline{\mathrm{Res}}_{\bar{z}\to\bar{z}_0}\tilde{j}(\bar{z})\mathcal{O}(z, \bar{z}_0) = \frac{1}{i\epsilon}\partial\mathcal{O}(z_0, \bar{z}_0). \tag{1.103}$$

This form of the Ward identity can be used to determine part of the form of the OPE of $T, \tilde{T}$ with an arbitrary operator. By considering scale transformations and translations one finds:

$$\begin{aligned} T(z)\mathcal{O}(w, \bar{w}) &\sim \cdots + \frac{1}{(z-w)^2}h\mathcal{O}(w, \bar{w}) + \frac{1}{z-w}\partial\mathcal{O}(w, \bar{w}), &(1.104)\\ \tilde{T}(\bar{z})\mathcal{O}(w, \bar{w}) &\sim \cdots + \frac{1}{(\bar{z}-\bar{w})^2}\tilde{h}\mathcal{O}(w, \bar{w}) + \frac{1}{\bar{z}-\bar{w}}\bar{\partial}\mathcal{O}(w, \bar{w}), &(1.105)\end{aligned}$$

where $(h, \tilde{h})$ is called the conformal weight of $\mathcal{O}$. A scale transformation is given by:

$$\mathcal{O}'(\zeta z, \bar{\zeta}\bar{z}) = \zeta^{-h}\bar{\zeta}^{-\tilde{h}}\mathcal{O}(z, \bar{z}). \tag{1.106}$$

As an example c^z has conformal weight (-1,0) and b_{zz} has weight (2,0). There is one further symmetry of (1.83) that comes in handy which is ghost number conservation. Its current is given by

$$j_{bc} = - : bc :, \quad \tilde{j}_{\tilde{b}\tilde{c}} = - : \tilde{b}\tilde{c} :, \tag{1.107}$$

so that c has (holomorphic) ghost number 1 and b has (holomorphic) ghost number -1.

As can be derived from (1.43) the BRST operator is given by

$$\begin{aligned} \mathcal{Q} &= Q + \tilde{Q}, &(1.108)\\ Q &= \int dz c(z)T^m(z) + \frac{1}{2} : c(z)T^g(z) : + a\partial^2 c(z), &(1.109)\\ \tilde{Q} &= \int d\bar{z}\tilde{c}(\bar{z})\tilde{T}^m(\bar{z}) + \frac{1}{2} : \tilde{c}(\bar{z})\tilde{T}^g(\bar{z}) : + a\bar{\partial}^2\tilde{c}(\bar{z}), &(1.110)\end{aligned}$$

The constant a is yet to be determined. Its value has no effect on the Noether charge Q since it multiplies a total derivative. To compute Q^2 it is convenient to write

$$Q = \int dz j_B(z). \tag{1.111}$$

Inside the path integral one has to use OPEs to evaluate Q^2:

$$Q^2 = \int dzdwj_B(z)j_B(w) = \int dzdw \sum_{i=-\infty}^{\infty} \mathcal{O}_i(w)(z-w)^i = \int dz\mathcal{O}_{-1}(z). \quad (1.112)$$

Hence a sufficient condition for quantum mechanic nilpotency of the BRST charge is vanishing of the single pole in the $j_B(z)j_B(w)$ OPE. This term can be calculated in a computation that requires some careful bookkeeping of minus signs and anticommuting variables:

$$j_B(z)j_B(w)|_{\frac{1}{z-w}} = \frac{1}{z-w}\left[(-\frac{3}{2}+a)\partial c\partial^2 c(w) + (\frac{d}{12} - a - \frac{2}{3})c\partial^3 c(w)\right], \quad (1.113)$$

where d is the number of X fields, in other words the dimension of spacetime. The solution to $Q^2 = 0$ is given by

$$a = \frac{3}{2}, \quad d = 26. \quad (1.114)$$

The second constraint has huge physical implications, it says the string theory defined by the Polyakov action is only a sensible quantum theory in 26 spacetime dimensions. Obviously this cannot describe nature. More physical string theories will be discussed in due course.

Closed string spectrum

Although the Polyakov string is not a realistic model it is still interesting to study its spectrum, because the lessons learnt from this exercise carry over to the more realistic string models. As follows from the main formula of Polyakov string theory, (1.83), vertex operators can be either integrated, U, or unintegrated, V:

$$V(z,\bar{z}) = c(z)\tilde{c}(\bar{z})V^m(z,\bar{z}), \quad U = \int d^2z V^m(z,\bar{z}), \quad (1.115)$$

where V^m only contains matter fields. Naively one would say V^m can be an arbitrary function of $X(z,\bar{z})$ before imposing $QV = 0$. However, for a well defined quantum theory it is necessary that all operators have single valued OPEs among each other. If one includes $X(z,\bar{z})$ in the spectrum of allowed operators, this rule is violated because the XX OPE is not single valued (cf. (1.94)). A good set of operators is given by

$$:\left(\prod_i \partial^{a_i} X^{m_i}(z)\right)\left(\prod_j \bar{\partial}^{b_j} X^{n_j}(\bar{z})\right) e^{ik\cdot X(z,\bar{z})}:, \quad (1.116)$$

where the argument of $\partial^k X$ is given as z because the equation of motion implies $\partial^k X$ is holomorphic, $\bar{\partial}\partial X = 0$. The weights of these operators are given by

$$h^{a_i} = \frac{\alpha' k^2}{4} + \sum_i a_i, \quad \tilde{h}^{b_j} = \frac{\alpha' k^2}{4} + \sum_j b_j. \quad (1.117)$$

The BRST procedure guarantees that

$$Q\int V^m = 0 \Leftrightarrow Q(cV^m) = 0. \tag{1.118}$$

So determining either the integrated or unintegrated cohomology is enough to obtain the spectrum. The physical state condition implies vanishing of the conformal weight of the vertex operator. The OPE of the BRST current with a vertex operator is given by:

$$j_B(z)c(w)V^m(w,\bar{w}) \sim \tag{1.119}$$

$$c(z)c(w)\left(\sum_{i=3}^{\infty}\frac{1}{(z-w)^i}\mathcal{O}^i(w,\bar{w}) + \frac{1}{(z-w)^2}hV^m(w,\bar{w}) + \frac{1}{z-w}\partial V^m(w,\bar{w})\right)$$

$$-\frac{1}{z-w}(\partial c)c(w)V^m(w,\bar{w}),$$

For certain operators $\mathcal{O}^i$, that depend on V^m and whose precise forms are irrelevant for the argument. The vanishing of the single pole in this OPE indeed implies vanishing of the conformal weight:

$$0 = j_B(z)c(w)V^m(w)|_{\frac{1}{z-w}} = \tag{1.120}$$

$$\frac{1}{z-w}\left[(h-1)(\partial c)c(w)V^m(w) + \sum_{k=2}^{\infty}\frac{1}{k!}(\partial^k c)c(w)\mathcal{O}^{k+1}(w,\bar{w})\right] \Rightarrow h-1=0.$$

Incidentally one can conclude that cV^m is only BRST closed if all $\mathcal{O}^k$ vanish, i.e. the OPE of the stress energy tensor and V^m does not have poles order three or higher. Similarly

$$\tilde{h} - 1 = 0. \tag{1.121}$$

Note $(h-1, \tilde{h}-1)$ is the conformal weight of the total vertex operator $c\bar{c}V^m$.

The quantity k in the vertex operators has the interpretation of physical spacetime momentum as can be inferred by constructing a Noether charge associated to spacetime translation invariance of the Polyakov action. The mass of the state is therefore given by $mass^2 = -k^2$. The constraint on the conformal weight can be transformed into a mass formula:

$$mass^2 = -k^2 = -\frac{4}{\alpha'} + \sum_i a^i = -\frac{4}{\alpha'} + \sum_j b^i \tag{1.122}$$

Incidentally the above formula contains a level matching condition, i.e. the holomorphic weight must be equal to the antiholomorphic weight. The mass spectrum is discrete and bounded from below. The first two mass levels are described below.

- $mass^2 = -\frac{4}{\alpha'}$ There is only one $\mathcal{Q}$ closed vertex operator at this mass level:

$$V(z,\bar{z}) = c(z)\tilde{c}(\bar{z})e^{ik\cdot X(z,\bar{z})}. \tag{1.123}$$

Furthermore it cannot be written as $V = \mathcal{Q}\Omega$ for the following reason. By ghost number conservation the ghost number of Ω must be 0. The conformal weight of Ω must vanish too, because Q does not change the weight. This narrows the space of possible Ω's down to operators of the form

$$\Omega = (bc)^n(\tilde{b}\tilde{c})^{\tilde{n}}V^m, \tag{1.124}$$

where the weight of the matter part must be $(-n, -\tilde{n})$. After some algebra one can deduce that none of these operators satisfies $\mathcal{Q}\Omega = V$.

From the spacetime point of view this is a Lorentz scalar (i.e. a spin zero particle). Since it has negative mass squared it is also a tachyon and this causes uncontrollable divergences when one considers scattering amplitudes. The tachyon is an artifact of Polyakov string theory and the more sophisticated models, to be described in due course, do not contain a tachyonic mode.

- $mass^2 = 0$ The BRST closed vertex operators at this mass level are given by

$$c(z)\tilde{c}(\bar{z})e_{mn}\partial X^m(z)\bar{\partial}X^n(\bar{z})e^{ik\cdot X(z,\bar{z})}, \tag{1.125}$$

subject to the constraints

$$k^2 = 0, \quad k^m e_{mn} = k^n e_{mn} = 0. \tag{1.126}$$

The exact states are given by

$$(Q+\tilde{Q})(f_m c(z)\partial X^m(z) + \tilde{f}_n\tilde{c}(\bar{z})\bar{\partial}X^n(\bar{z}))e^{ik\cdot X(z,\bar{z})} = \tag{1.127}$$

$$c(z)\tilde{c}(\bar{z})(k_m\tilde{f}_n + f_m k_n)\partial X^m(z)\bar{\partial}X^n(\bar{z})e^{ik\cdot X(z,\bar{z})}, \quad k^2 = 0, \quad f\cdot k = \tilde{f}\cdot k = 0,$$

where the conditions on k, f and $\tilde{f}$ are crucially used in the equality. Hence on top of the constraint the polarisation e_{mn} also has a gauge invariance:

$$e_{mn} \cong e_{mn} + k_m\tilde{f}_n + f_m k_n, \quad f\cdot k = \tilde{f}\cdot k = 0. \tag{1.128}$$

In conclusion the massless excitations form a rank two Lorentz tensor and the physical excitations are the transversal ones. Within this rank two tensor there are three parts that never mix under Lorentz transformations, so all observers would agree that it is possible to divide the massless spectrum into three groups. The decomposition is given by

$$e^{mn} \to e^{(mn)} - \frac{1}{26}\eta^{mn}e^p_p \oplus e^{[mn]} \oplus e^m_m. \tag{1.129}$$

One recognises a symmetric traceless tensor which is interpreted as the graviton. This is natural because the constraints and gauge invariances on $e_{(mn)}$ are precisely those that one finds by plugging $g_{mn} = \eta_{mn} + e_{(mn)}$ in the Einstein Hilbert action. The antisymmetric part are the physical excitations of an antisymmetric rank two tensor. In flat space its action is given by

$$\int d^d x \partial_{[m} B_{np]} \partial^{[m} B^{np]}. \tag{1.130}$$

The massless scalar is called the dilaton.

This concludes the discussion on the closed string spectrum. The states with positive $mass^2$ are not of direct interest because their mass is of the order of the Planck scale. The logical next step would be to compute some scattering amplitudes but a number of features of the Polyakov string is very unphysical, e.g. 26 dimensions, the tachyon, no fermions. These will be dealt with first and amplitude computation will be discussed in more physical theories.

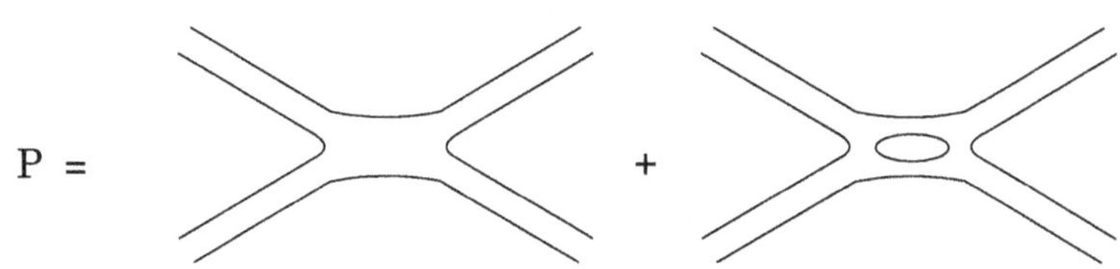

+ higher loops

Figure 1.7: *Scattering of open strings*

1.2.2 Open strings

The open string analog of figure 1.4 is depicted in figure 1.7. Due to the presence of a worldsheet boundary the open string Polyakov action does not have as much gauge invariance as its closed string analog. More explicitly the parameter of the gauge transformations in (1.18) is restricted by a boundary condition:

$$n_a v^a(\sigma) = 0, \quad \sigma \in \partial\Sigma, \tag{1.131}$$

where n_α is the normal vector to boundary $\partial\Sigma$. The above condition on v^α carries over to the BRST ghost c^α. The Weyl (or conformal) invariance is unchanged. This can be exploited to map the tree diagram in figure 1.7 to the disk as shown in figure 1.8, where the vertex operators are inserted at points on the boundary.

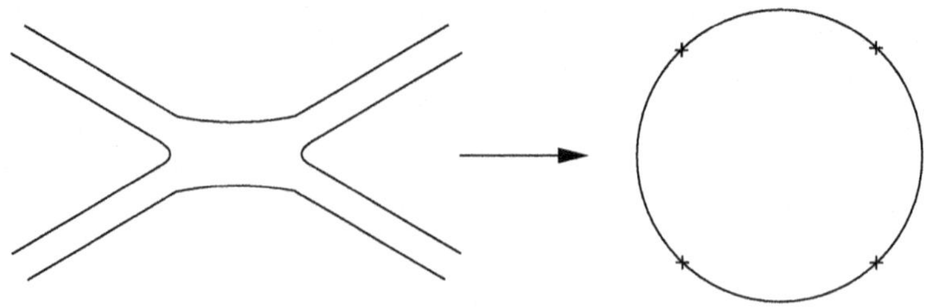

Figure 1.8: *Scattering of open strings; the four crosses represent vertex operators*

In turn the disk can be mapped by a conformal transformation to the upper half of the complex plane. On the upperhalf plane the boundary condition on c^α implies

$$c(z,\bar{z}) = \tilde{c}(z,\bar{z}), \quad \mathrm{Im} z = 0. \tag{1.132}$$

It is very useful to combine these two fields, one holomorphic and one antiholomorphic with both only defined on the upperhalf plane, into one holomorphic field that is defined on the whole complex plane:

$$c(z,\bar{z}) \equiv \tilde{c}(\bar{z},z), \quad \mathrm{Im} z < 0. \tag{1.133}$$

There is a similar boundary condition on the b ghost. This results from the fact that the variation of an action on a manifold with a boundary also contains boundary terms. In case of the Faddeev Popov action (1.82) this becomes

$$\int_{\partial\Sigma} ds n^a b_{ab} \delta c^b. \tag{1.134}$$

This has to vanish and therefore

$$n_a t_b b^{ab} = 0, \tag{1.135}$$

where t^a is a tangent vector to the boundary. On the upperhalf plane the above condition reads

$$b(z,\bar{z}) = \tilde{b}(z,\bar{z}), \quad \mathrm{Im} z = 0. \tag{1.136}$$

So that

$$b(z,\bar{z}) \equiv \tilde{b}(\bar{z},z), \quad \mathrm{Im} z < 0. \tag{1.137}$$

combines the two b ghosts into one holomorphic one. The stress energy tensor on a surface with a boundary also satisfies a boundary condition:

$$t^a n^b T_{ab} = 0. \tag{1.138}$$

In order to see this first note since the translation invariance is partially broken, only $T_{ab}t^b$ is a conserved current anymore. Equation (1.138) is then the statement the conserved current cannot flow out of the manifold. The stress energy tensor can now be treated in the same way as b:

$$T(z,\bar{z}) \equiv \tilde{T}(\bar{z},z), \quad \mathrm{Im}z < 0. \tag{1.139}$$

With this definition the holomorphic object T encapsulates all information of the stress energy tensor. Also note that this doubling trick can also be applied to the BRST current since it (only) contains fields discussed just above.

The BRST operator of open string theory, which is the sum of an open holomorphic and an open antiholomorphic contour integral in the upperhalf plane, can be written as a closed holomorphic integral in the complex plane (cf. figure 1.9):

$$\mathcal{Q} = Q + \tilde{Q} = \int dz j_B(z) + \int d\bar{z}\tilde{j}_B(\bar{z}) = \oint dz j_B(z). \tag{1.140}$$

The spectrum of open string theory can be obtained by studying the cohomology

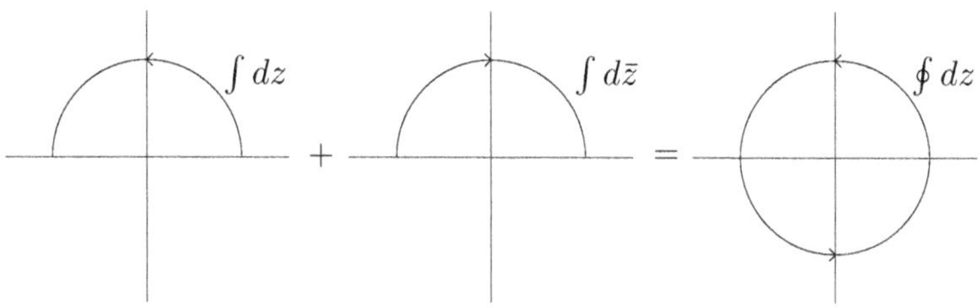

Figure 1.9: *Contour integrals*

of the BRST operator, which is very much like half of the closed string discussion. A good set of (boundary) operators is given by

$$\vdots(\partial_y)^k X^m(y) e^{ik\cdot X(y)}\vdots, \tag{1.141}$$

where y parametrises the boundary and $\vdots\mathcal{O}(y)\vdots$ denotes boundary normal ordering. Its explicit form is not needed in this thesis but can be derived by going through the steps leading to (1.94) for manifolds with a boundary. The mass levels of the open string are given by

$$(mass)^2 = \frac{-1+k}{\alpha'}. \tag{1.142}$$

The first two mass levels are described below.

- $mass^2 = -\frac{1}{\alpha'}$ This is a tachyon, its vertex operator is given by

$$V(y) = c(y)\vdots e^{ik\cdot X(y)}\vdots. \tag{1.143}$$

- $mass^2 = 0$ The massless states constitute the physical excitations of a photon:

$$V(y) = e_m c(y) \vdots (\partial_y) X^m(y) e^{ik \cdot X(y)} \vdots, \quad k_m e^m = 0, \quad e_m \cong e_m + k_m. \tag{1.144}$$

For completeness the main formula of open string theory is specified below:

$$Z = \int d^{\frac{\mu}{2}} \tau \mathcal{D}X \mathcal{D}b \mathcal{D}c d^{N-3} y^i \prod_{i=1}^{3} \sqrt{\hat{g}(\hat{y}^i)} c(\hat{y}^i) V_i(\hat{y}^i) \prod_{j=4}^{N} \sqrt{\hat{g}(y^j)} V_j(y^j)$$

$$\prod_{k=1}^{\frac{\mu}{2}} (b, \partial_k \hat{g}) e^{\frac{-1}{2\pi\alpha'} \int_\Sigma d^2\sigma \sqrt{\hat{g}} \hat{g}^{ab} \partial_a X^\mu \partial_b X_\mu - (b, \hat{P}_1 c)}. \tag{1.145}$$

1.2.3 Curved backgrounds

As discussed above the Polyakov only defines a consistent quantum theory in 26 dimensions. Moreover string theory has been defined above as the infinite sum of diagrams as in figure 1.4, but there is no small parameter associated to them such that the higher loops become negligible when the parameter is small enough. This section introduces a generalisation of the worldsheet action that resolves these issues. The choice for the Polyakov action, (1.16), for the weight in the path integral was based on simplicity. This is not a very good principle. A better strategy is writing down the most general worldsheet action for the embedding coordinates X with at most two worldsheet derivatives:

$$S_\sigma = \frac{1}{4\pi\alpha'} \int d^2\sigma \sqrt{g} \left[\left(g^{\alpha\beta} G_{\mu\nu}(X) + \epsilon^{\alpha\beta} B_{\mu\nu}(X) \right) \partial_\alpha X^\mu \partial_\beta X^\nu + \alpha' R^{(2)} \Phi(X) \right]. \tag{1.146}$$

The field $G_{\mu\nu}$ is symmetric and has the interpretation of the background spacetime metric. This is the vacuum metric about which quantum modes can get excited. The field $B_{\mu\nu}$ is antisymmetric, it will become clear this field is intimately related with the antisymmetric modes in the massless closed string spectrum. The last term contains a spacetime scalar which is the background value of the dilaton. The α' has to be included for the spacetime dimensions to work out.

A priori one can try to compute amplitudes for arbitrary choices of the background fields. For example the Polyakov action was defined by $G_{\mu\nu} = \eta_{\mu\nu}, B_{\mu\nu} = 0, \Phi = 0$. Already for this choice anomaly cancellation (i.e. quantum nilpotency of the BRST operator) imposed restrictions on the dimension of spacetime. After a short discussion on the dilaton it will be shown that anomaly cancellation of the theory defined by S_σ implies very natural equations on the background fields. The lack of a small parameter problem can be resolved by looking at a background with a constant dilaton:

$$\Phi(X) = \Phi_0. \tag{1.147}$$

For a constant dilaton every diagram in the string loop expansion (figure 1.4) gets multiplied by a factor

$$e^{-S_{\Phi_0}}, \quad S_{\Phi_0} = \frac{1}{4\pi}\int_{\Sigma_g} d^2\sigma\sqrt{g}R^{(2)}\Phi_0 \tag{1.148}$$

This is a topological invariant, i.e. it only depends on quantities like the genus and the number of boundary components of the worldsheet but not on the metric. The integral can be evaluated as

$$e^{-S_{\Phi_0}} = (g_s)^{2-2g-b}, \quad g_s = e^{-\Phi_0}, \tag{1.149}$$

where g is the genus of the worldsheet and b the number of boundary components, e.g. $(g,b) = (0,2)$ is the cylinder. When g_s is small it is now possible to approximate the amplitude by the first few diagrams.

In the discussion of the vertex operators, overall normalisations were not included. These might appear to be free parameters of the theory. It turns out however that all the normalisation factors are related to each other and the string coupling g if one imposes constraints that follow from unitarity of the S matrix. In this thesis these normalisations will be suppressed.

The massless spectrum of the Polyakov string theory contained the physical (i.e. transverse) excitations of a graviton and a photon. The condition $k \cdot e = 0$ and the gauge invariance $e \cong e + k$ were consequences of the fact that the Weyl symmetry was a quantum symmetry. The first two terms in the generalised action in (1.146) are classically invariant under Weyl transformations. In other words the trace of the stress energy tensor vanishes if the equations of motion are used. The third term is only invariant under rigid Weyl transformation. However, it is still possible to find background fields such that correlators are Weyl invariant. For arbitrary background fields the Weyl anomaly is given by

$$\delta_W \int \mathcal{D}\phi \mathcal{O}_1 \cdots \mathcal{O}_N e^{-S_\sigma(g)} = \int d^2\sigma\sqrt{g}\delta\omega(\sigma)\int \mathcal{D}\phi T^a_a(\sigma)\mathcal{O}_1\cdots\mathcal{O}_N e^{-S_\sigma}, \tag{1.150}$$

where

$$T^a_a = -\frac{1}{2\alpha'}\beta^G_{\mu\nu}g^{ab}\partial_a X^\mu \partial_b X^\nu - \frac{i}{2\alpha'}\beta^B_{\mu\nu}\epsilon^{ab}\partial_a X^\mu\partial_b X^\nu - \frac{1}{2}\beta^\Phi R^{(2)}. \tag{1.151}$$

The explicit expressions for the β functions can be obtained by expanding the background fields about a background. The details of the computations can be found in [2]. The answer is given by

$$\beta^G_{mn} = \alpha' R^{(d)}_{mn} + 2\alpha'\nabla_m\nabla_n\Phi - \frac{\alpha'}{4}H_{mpr}H_n{}^{pr} + O(\alpha'^2), \tag{1.152}$$

$$\beta^B_{mn} = -\frac{\alpha'}{2}\nabla^p H_{pmn} + \alpha'\nabla^p\Phi H_{pmn} + O(\alpha'^2), \tag{1.153}$$

$$\beta^\Phi = \frac{D-26}{6} - \frac{\alpha'}{2}\nabla^2\Phi + \alpha'\nabla_p\Phi\nabla^p\Phi - \frac{\alpha'}{24}H_{mnp}H^{mnp} + O(\alpha'^2), \tag{1.154}$$

where H_{mnp} is the three form field strength of the two form gauge field B_{mn} defined by

$$H_{mnp} = \partial_{[m}B_{np]} \tag{1.155}$$

Note that H is invariant under gauge transformations $B_{mn} \rightarrow B_{mn} + \partial_{[m}\zeta_{n]}$. Quantum Weyl invariance now translates to

$$\beta^G_{mn} = \beta^B_{mn} = \beta^\phi = 0. \tag{1.156}$$

The vanishing of β^G_{mn} is precisely Einsteins equations for the metric in the presence of sources, in the form of a two form gauge field and a scalar, the dilaton. The vanishing of β^B_{mn} gives the field equation for the two form gauge field and the last β function is the equation of motion of a scalar coupled to gravity and a two form gauge field. In conclusion string theory in curved backgrounds only leads to a sensible theory if the background satisfies the classical field equations (1.152)-(1.154).

The action (1.146) paves the way for more realistic string theories. One can for instance choose G_{mn} to be the metric of a space of the form $\mathcal{M}^4 \times M_{22}$, i.e. four dimensional Minkowski space times a 22 dimensional compact space. When this space is chosen to be T^{22}, i.e. the internal space is a torus, the closed string spectrum contains a tachyon with $mass^2 \sim -\frac{\alpha'}{R}$, where R is the radius of the torus. The vertex operators of the massless states are given by

$$\partial X^m \bar{\partial} X^n e^{ik\cdot X}, \quad \left(\partial X^m \bar{\partial} X^i + \partial X^i \bar{\partial} X^m\right) e^{ik\cdot X}, \tag{1.157}$$

$$\left(\partial X^m \bar{\partial} X^i - \partial X^i \bar{\partial} X^m\right) e^{ik\cdot X}, \quad \partial X^i \bar{\partial} X^j e^{ik\cdot X}, \quad m,n = 0,1,2,3, \quad i,j = 4,\ldots,25.$$

Note the above operators are subject to modding out by BRST exact states. The physical excitations of the first vertex operator contains the (4d) spacetime graviton, antisymmetric tensor and dilaton. The second one contains 22 spacetime vectors, which are standard Kaluza Klein modes that come about when dimensionally reducing. The second line contains the Kaluza Klein vectors from the antisymmetric tensor and 22^2 scalars, which can be thought of as moduli (or flat directions) of the metric and the antisymmetric tensor. The next mass level is at $mass^2 \sim \frac{\alpha'}{R}$. Hence if the radius of the torus is small these states will be very massive and unobservable by experiments at typical particle accelerator energies.

The above proposal for the internal manifold is a step in the right direction, because one finds a four dimensional spectrum. However, the existence of huge number of massless scalars is worrisome, because these are not observed in nature. One needs to look for more sophisticated choices of the internal manifold. This direction will not be pursued in this thesis. Instead a string theory will be constructed whose spectrum on the one hand does not have a tachyon and on the other hand includes fermions.

1.3 RNS formalism and its limitations

One way of introducing fermions into the theory is making it supersymmetric. As mentioned in the beginning of this chapter there are physical reasons to look for supersymmetric theories. Furthermore the inclusion of extra symmetries provides a powerful tool to do computations. A spacetime supersymmetric string theory, which goes by the name of RNS, is defined by the worldsheet action obtained by covariantising the Polyakov action under $\mathcal{N}=1$ worldsheet supersymmetry [7, 8]:

$$S_{RNS} = \frac{1}{2\pi\alpha'}\int d^2\sigma\sqrt{g}[\frac{1}{2}g^{ab}\partial_a X^m\partial_b X_m + \frac{1}{2}i\psi^m\gamma^a\nabla_a\psi_m + \frac{1}{2}i(\chi_a\gamma^b\gamma^a\psi_m)(\partial_b X^m - \frac{1}{4}i\chi_b\psi^m)], \tag{1.158}$$

where worldsheet fermion indices are suppressed. The fields X^m are the usual spacetime coordinates and ψ^m are their 2d supersymmetric partners. Moreover ψ^m are Majorana spinors, so that these fields have two real components each in two dimensions. The field g_{ab} is the worldsheet metric and χ_a is the worldsheet gravitino. Spacetime supersymmetry is present but not manifest from the worldsheet point of view. Quantisation of the RNS string works in a similar fashion to the bosonic string. In particular there will be a BRST operator. In the RNS formalism, however, this operator is only nilpotent in ten spacetime dimensions. Therefore in the sequel of this thesis spacetime will be assumed to be ten dimensional.

Symmetries and gauge fixing

The above action is invariant under diffeomorphisms and local 2d $\mathcal{N}=1$ supersymmetry, the latter is given by

$$\delta g_{ab} = 2i\epsilon\gamma_{(a}\chi_{b)}, \tag{1.159}$$
$$\delta\chi_a = 2\nabla_a\epsilon, \tag{1.160}$$
$$\delta X^m = i\epsilon\psi^m, \tag{1.161}$$
$$\delta\psi^m = \gamma^a(\partial_a X^m - \frac{1}{2}i\chi_a\psi^m)\epsilon. \tag{1.162}$$

Moreover (1.158) is invariant under Weyl and super Weyl transformations [9]:

$$X^m \to X^m, \quad g_{ab}\to e^{2\omega}g_{ab}, \quad \psi^m \to e^{\omega}\psi^m, \quad \chi_a \to e^{-\omega}\chi_a + \gamma_a\lambda, \tag{1.163}$$

where λ is the parameter of the super Weyl transformations.

The presence of worldsheet fermions makes quantisation much more complicated. In particular the moduli space is no longer an ordinary finite dimensional manifold, but also includes anticommuting variables. In other words moduli space becomes a supermanifold. Furthermore there will be ghosts for the local supersymmetry.

The particular form of the insertions containing these new fields is hard to derive. Therefore quantisation of the RNS string on the sphere, where there is no moduli space, is discussed first.

1.3.1 Tree-level

On the sphere the gauge invariances can be fixed by

$$g_{ab} = \hat{g}_{ab} = e^{2\omega_0}\delta_{ab}, \quad \chi_a = \hat{\chi}_a = \gamma_a\lambda_0. \tag{1.164}$$

The gauge fixed action will not depend on ω_0 and λ_0. A convenient choice is to set them both to zero. In this gauge, the superconformal gauge, the worldsheet action reads

$$S_{X\psi} = \int d^2z \left(\partial X^m \bar{\partial} X_m + \psi_m \bar{\partial}\psi^m + \tilde{\psi}_m \partial\tilde{\psi}^m\right), \tag{1.165}$$

where ψ^m and $\tilde{\psi}^m$ are the two components of the 2d spinor. In closed string theory the X coordinates are periodic. In the z coordinate, the one defined on the whole complex plane, this condition can be expressed as the statement that $X^m(ze^{i\alpha})$ is periodic in α, with period 2π. Note, however, the action is also well defined for antiperiodic boundary conditions on X^m, i.e. $X^m(ze^{i\alpha})$ is antiperiodic in α. This can lead to unitary string theories, but this direction will not be pursued in this thesis. On the other hand antiperiodic boundary conditions on the fermion will turn out to be of vital importance. The two possible boundary conditions have a name:

$$Neveu-Schwarz \;:\; \psi^m(e^{2\pi i}z) = +\psi^m(z), \tag{1.166}$$

$$Ramond \;:\; \psi^m(e^{2\pi i}z) = -\psi^m(z). \tag{1.167}$$

Together with the boundary conditions for the antiholomorphic side this leads to four sectors: (NS,NS), (NS,R), (R,NS) and (R,R). From (1.167) one sees that the ψ part of a Ramond sector vertex operator is double valued. Such a function can pictorially be represented by the end of a branchcut, cf. figure 1.10. This figure also shows that the number of R states must be even, otherwise the branchcuts have nowhere to go. In a later paragraph the R sector will be related to spacetime fermions. The number of these fields must also always be even in scattering amplitudes.

Ghost action

The ghost action for the RNS string contains an extra pair of ghost fields that are related to the local supersymmetry. Since the parameter of this symmetry is fermionic, the ghost fields are bosonic and they are denoted β, γ. To write down the action for these new ghost fields, it is useful to note the operators P_n and P_n^T can

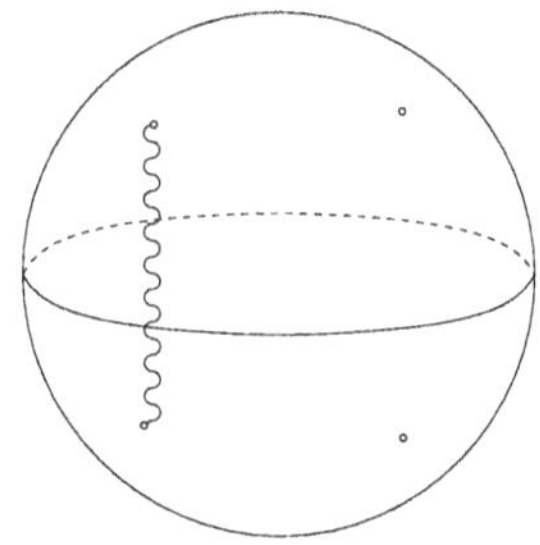

Figure 1.10: *Tree level four point function with two (R,NS) sector states (left) and two (NS,NS) sector states (right)*

also be defined on fields of half integer spin, i.e. spinors. For details see [3]. The supersymmetry transformation of the gravitino (1.160) can be written as

$$\delta\chi_a = (P^{1/2}\epsilon)_a. \tag{1.168}$$

When the statistics of the fermionic parameter ϵ is flipped, it becomes the Faddeev-Popov ghost γ. The other ghost field, β, has indices such that it can be contracted with χ_a, i.e. it has spin $\frac{3}{2}$. This leads to the following ghost action:

$$S_{\beta\gamma} = \int d^2\sigma\sqrt{g}\beta_a(P^{1/2}\gamma)^a + \cdots, \tag{1.169}$$

where the ellipsis contains the gravitino. In superconformal gauge this ghost action can be written as

$$S_{\beta\gamma c.g.} = \int d^2z\left(\beta\bar{\partial}\gamma + \tilde{\beta}\partial\tilde{\gamma}\right), \tag{1.170}$$

where β and $\tilde{\beta}$ have conformal weights $(\frac{3}{2},0)$ and $(0,\frac{3}{2})$ respectively, γ and $\tilde{\gamma}$ have weights $(-\frac{1}{2},0)$ and $(0,-\frac{1}{2})$ respectively.

As in the bosonic string the gauge condition (1.164) does not completely fix all invariances. The CKVs encountered in section 1.2 remain for the RNS string. In addition there are super conformal Killing vectors (SCKVs), which represent the residual local supersymmetry. Some vertex operators will contain a factor of $\delta(\gamma)$. (cf. the bosonic string where the Faddeev Popov ghosts are fermionic, hence $\delta(c) = c$.) Quantising the worldsheet action of the RNS formalism at an arbitrary genus, which includes finding the precise form of these insertions, is a formidable task due to the complicated nature of the (super)moduli spaces, cf. [3]. On the sphere, however, matters simplify and this case is described below.

Charge conservation in $\beta\gamma$ systems

One can use the charge conservation anomaly to deduce correlators are only non vanishing if the ghost number has a certain value, depending on the conformal weight of the ghosts. Explicitly, consider the action:

$$\int d^2z\beta\bar{\partial}\gamma, \tag{1.171}$$

where β and γ are conjugate fields of weight λ and $1-\lambda$ respectively. They can be either bosonic ($\epsilon=-1$) or fermionic ($\epsilon=+1$). This paragraph is partly based on [8]. The OPEs are given by

$$\gamma(z)\beta(w)\sim\frac{1}{z-w},\quad \beta(z)\gamma(w)\sim\frac{\epsilon}{z-w}. \tag{1.172}$$

The action is invariant under translations, it also possesses a $U(1)$ symmetry called ghost number. The generators are given by:

$$T(z)=-\lambda\beta\partial\gamma(z)+(1-\lambda)(\partial\beta)\gamma(z),\quad j(z)=-\beta\gamma(z), \tag{1.173}$$

where coincident operators are understood to be normal ordered. This convention of suppressing the colons will also be used in the sequel of this thesis. The OPE of these two operators reads

$$T(z)j(w)\sim\frac{\epsilon(1-2\lambda)}{(z-w)^3}+j(w)\frac{1}{(z-w)^2}+\partial j(w)\frac{1}{z-w}. \tag{1.174}$$

In order to show only operators of a certain ghost number are non vanishing consider

$$\langle N^g\mathcal{O}\rangle, \tag{1.175}$$

where the number operator is given by:

$$N^g=\oint\frac{dz}{2\pi i}j(z) \tag{1.176}$$

and $\mathcal{O}$ is an arbitrary operator of ghost number $Q_{\mathcal{O}}$, i.e. $[N^g,\mathcal{O}]=Q_{\mathcal{O}}\mathcal{O}$. The correlator (1.175) can be evaluated in two ways, either by pulling the contour off of the back of the sphere or by shrinking it in the z patch passing through all insertions. The latter gives

$$Q_{\mathcal{O}}\langle\mathcal{O}\rangle. \tag{1.177}$$

To evaluate the former one needs to rewrite N^g in the u patch. The transformation can be derived from Tj OPE with the help of the Ward identity (1.102) (where $\mathcal{O}=-\beta\gamma$ and $j(z)=v(z)T(z)$):

$$\frac{1}{\epsilon}\delta j=-v\partial j-j\partial v+\epsilon\frac{2\lambda-1}{2}\partial^2 v. \tag{1.178}$$

Note j denotes the ghost number current. The finite form of the above transformation is given by

$$(\partial_z u)j_u(u)=j_z(z)+\epsilon\frac{2\lambda-1}{2}\frac{\partial_z^2 u}{\partial_z u}. \tag{1.179}$$

Using this result one can rewrite N^g in the u patch:

$$\oint \frac{dz}{2\pi i} j(z) = \oint \frac{1}{2\pi i}\frac{du}{u^2}[j_u(u)u^2 - \epsilon(2\lambda-1)u] = -\epsilon(2\lambda-1) + \oint \frac{du}{2\pi i} j_u(u) \quad (1.180)$$

and since there are no insertions inside the contour anymore the contour integral vanishes, hence

$$\langle \oint \frac{dz}{2\pi i} j(z)\mathcal{O}\rangle = -\epsilon(2\lambda-1)\langle\mathcal{O}\rangle \quad (1.181)$$

In conclusion, if $-Q_{\mathcal{O}} \neq \epsilon(2\lambda-1)$ then $\langle\mathcal{O}\rangle = 0$.

This agrees with the results of the bosonic string, because $\epsilon(2\lambda-1)$ is minus three in that case. Hence only operators with ghost number three are non vanishing. All tree level amplitudes have this property. For the bosonic ghosts of the RNS string the above argument implies the total charge of the γ insertions must be minus two. Note this can be achieved by two $\delta(\gamma)$'s because $[N^g, \delta(\gamma)] = -\delta(\gamma)$. Since $\delta(\gamma)$ is not written as a proper function of worldsheet fields it is difficult to perform computations. In the next paragraph $\delta(\gamma)$ will be given in terms of a smooth function of fields of a different worldsheet CFT. The precise map between the two CFTs will be given. Furthermore this mapping allows for writing down the double valued vertex operator, needed for the R sector states.

Bosonisation

Bosonisation is a map from a given CFT to another CFT that respects all OPEs. In case of the CFT formed by two of the ten holomorphic fields $\psi^m(z)$, say ψ^1 and ψ^2, the OPEs between the the ψ's themselves can be given as

$$\psi(z)\bar{\psi}(w) \sim \frac{1}{z-w}, \quad \psi(z)\psi(w) \sim 0, \quad \bar{\psi}(z)\bar{\psi}(w) \sim 0, \quad (1.182)$$

where

$$\psi(z) = \frac{1}{\sqrt{2}}(\psi^1(z) + i\psi^2(z)), \quad \bar{\psi}(z) = \frac{1}{\sqrt{2}}(\psi^1(z) - i\psi^2(z)). \quad (1.183)$$

The second, equivalent, CFT consists of the holomorphic part of a boson, i.e. it satisfies the OPE:

$$H(z)H(w) \sim -\ln(z-w). \quad (1.184)$$

The map between the two CFTs is given by

$$\psi(z) \cong e^{iH(z)}, \quad \bar{\psi}(z) \cong e^{-iH(z)}. \quad (1.185)$$

The form of the number current and the stress energy tensor in the H CFT can be obtained via the OPEs:

$$e^{iH(z)}e^{-iH(w)} = \frac{1}{z-w} + i\partial H(w) + \partial H\partial H(w)(z-w) + O((z-w)^2), \quad (1.186)$$

$$\psi(z)\bar{\psi}(w) = \frac{1}{z-w} + \psi\bar{\psi}(w) - \psi\partial\bar{\psi}(w)(z-w) + O((z-w)^2), \quad (1.187)$$

where one recognises the number currents in the $O(1)$ terms and the stress energy tensors in the next ones. Note $e^{iH(z)}$ is single valued since $H(z)$ is single valued. An obvious realisation of a double valued operator constructed from $H(z)$ is given by

$$e^{isH(z)}, \quad s \in \mathbb{Z} + \frac{1}{2}. \tag{1.188}$$

Indeed the ψ part of the R state vertex operators has the above form. In the actual case there are five $\psi\bar{\psi}$ pairs, so there will be five copies of H, which are labelled by $p = 1, \ldots, 5$: H^p. By carefully studying the commutation relations of the R sector vertex operators, one can discover the need for cocycles [10]. These are exponentials of the zero modes of H. However these cocycles only affect relative signs of certain amplitudes and one can often ignore them. The weight of $e^{ilH(z)}$ can be obtained from its OPE with the stress energy tensor. The answer is $h = \frac{1}{2}l^2$.

An explicit form of $\delta(\gamma)$ can be found be performing a similar mapping for the bosonic $\beta\gamma$ CFT with $\lambda = \frac{3}{2}$. The equivalent CFT consists of the chiral bosons:

$$\phi(z)\phi(w) \sim -\ln(z-w), \quad \chi(z)\chi(w) \sim \ln(z-w). \tag{1.189}$$

The mapping is given by

$$\beta \cong e^{-\phi+\chi}\partial\chi, \quad \gamma \cong e^{\phi-\chi}. \tag{1.190}$$

The number current and the stress energy tensor read

$$\beta\gamma \cong \partial\phi, \quad T \cong -\frac{1}{2}\partial\phi\partial\phi + \frac{1}{2}\partial\chi\partial\chi - \partial^2\phi + \frac{1}{2}\partial^2\chi. \tag{1.191}$$

The operator $\delta(\gamma)$ should have ghost number minus one and obey the OPE

$$\delta(\gamma)(z)\gamma(w) = O(z-w). \tag{1.192}$$

These two conditions have a solution:

$$\delta(\gamma) \cong e^{-\phi}. \tag{1.193}$$

This concludes the discussion on bosonisation. The results of this paragraph will be used when the vertex operators are constructed.

Spectrum

The spectrum of physical states is as usual given by the cohomology of the BRST operator:

$$Q = \frac{1}{2\pi i}\oint \left(dz j_B - d\bar{z}\tilde{j}_B\right). \tag{1.194}$$

The BRST current is, up to total derivatives, given by

$$j_B = cT_{mat} + \gamma G_{mat} + \frac{1}{2}\left(cT_{gh} + \gamma G_{gh}\right) \tag{1.195}$$

where T_{mat} is the stress energy tensor of the matter sector which now includes a term containing ψ's, T_{gh} is the stress energy tensor of the ghost fields, which can be read off from (1.173). The current associated to worldsheet supersymmetry is denoted G, this also splits up in a matter and a ghost part. They are given by

$$T_{mat} = -\frac{1}{\alpha'}\partial X^\mu \partial X_\mu - \frac{1}{2}\psi^m \partial \psi_m, \quad T_{gh} = -2b\partial c - (\partial b)c - \frac{3}{2}\beta\partial\gamma - \frac{1}{2}(\partial\beta)\gamma, \tag{1.196}$$

$$G_{mat}(z) = i\sqrt{\frac{2}{\alpha'}}\psi^m \partial X_m(z), \quad G_{gh}(z) = (\partial\beta)c(z) + \frac{3}{2}\beta\partial c(z) - 2b\gamma(z). \tag{1.197}$$

The spectrum is worked out in [2]. Let us start by looking at only the holomorphic part of the vertex operator and start with the NS sector, which is characterised by the singlevaluedness of the ψ and $\beta\gamma$ sector vertex operators. Similar to the bosonic string the mass spectrum is discrete and bounded from below. The first two levels are given below.

- $mass^2 = -\frac{1}{2\alpha'}$ The (holomorphic part of the) $\mathcal{Q}$ closed operator at this mass level is given by

  $$V_T = c(z)e^{-\phi(z)}e^{ik\cdot X(z,\bar{z})}, \quad k^2 = -\frac{1}{2\alpha'}. \tag{1.198}$$

  Note $e^{-\phi}$ comprises the $\beta\gamma$ part of the vertex operator and it is indeed singlevalued. This is a tachyon. However in the RNS formalism there is a consistent way of removing this mode from the spectrum, which is described below.

- $mass^2 = 0$ The BRST closed vertex operators at this mass level are given by

  $$V_{NS} = e_\mu c(z)\epsilon^{-\phi(z)}\psi^\mu(z)e^{ik\cdot X(z\,\bar{z})}, \quad k^2 = 0, \quad k\cdot e = 0, \quad e \cong e + k. \tag{1.199}$$

  Hence the physical content is a massless vector. In open string theory this would be a photon. In closed string theory this could be part of the graviton, that is when the antiholomorphic side is also a massless vector.

These are all the NS sector states with non positive mass squared. In the Ramond sector all physical states have non negative mass squared. Recall that the ψ part of Ramond sector states is doublevalued. The massless states are given below.

- $mass^2 = 0$ The BRST closed vertex operators are given by

  $$V_R = \sum_{s_p=\pm\frac{1}{2}} u_{\mathbf{s}}c(z)e^{-\phi(z)}e^{is_pH^p}, \quad k^2 = 0, \quad k^\mu \Gamma^\mu_{\mathbf{s},\mathbf{s}'}u_{\mathbf{s}'} = 0, \tag{1.200}$$

where $\mathbf{s} = (s_1, s_2, s_3, s_4, s_5)$ and as indicated under the sum all s_p are either $1/2$ or $-1/2$ so that $\mathbf{s}$ takes 32 values. Both the ψ and the $\beta\gamma$ part are indeed doublevalued, since they involve square roots of holomorphic operators. The polarisation spinor, $u_{\mathbf{s}}$, has 32 components to start with. This is a Dirac spinor, which is a reducible representation that consists of two 16 component Weyl spinors of opposite chirality[4]. The physical state condition removes eight components of each Weyl spinor. In conclusion the spectrum consists of the physical modes of two Weyl spinors with opposite chirality.

A consistent string theory cannot contain all of the above states. An important consistency condition is that the OPEs of all vertex operators do not contain branch cuts. Furthermore tachyons cannot be present in any physical theory. Luckily these two conditions are compatible. A consistent string theory is obtained by removing the tachyon from the NS sector and one of the two Weyl spinors from the Ramond sector. For closed string theory this basically leaves two possibilities. Either the chirality of the Weyl spinor on the holomorphic side has the same chirality as the Weyl spinor on the antiholomorphic side or the opposite. The former is called type IIB string theory, the latter type IIA and the projection is known as the GSO projection. The spectrum of the two theories is summarised in table 1.2. The physical states are obtained by applying the appropriate field equation, e.g. for a p-form field strength this is $k_{m_1} f_{(p)}^{m_1 \cdots m_p} = 0$.

	IIB	IIA
NS-NS	$\phi, B_{[mn]}, h_{(mn)}$ (traceless)	$\phi, B_{[mn]}, h_{(mn)}$ (traceless)
NS-R	$(\lambda_1)^\alpha, (\psi_1)_{\alpha m}$ $(\gamma_m^{\alpha\beta}(\psi_1)_\beta^m = 0)$	$(\lambda_1)_\alpha, (\psi_1)^{\alpha m}$ $(\gamma^m_{\alpha\beta}(\psi_1)^\beta_m = 0)$
R-NS	$(\lambda_2)^\alpha, (\psi_2)_{\alpha m}$ $(\gamma_m^{\alpha\beta}(\psi_2)_\beta^m = 0)$	$(\lambda_2)^\alpha, (\psi_2)_{\alpha m}$ $(\gamma_m^{\alpha\beta}(\psi_2)_\beta^m = 0)$
R-R	$f_{(1)}^m, f_{(3)}^{[m_1 m_2 m_3]}, f_{(5)}^{[m_1 m_2 m_3 m_4 m_5]}$	$f_{(2)}^{[m_1 m_2]}, f_{(4)}^{[m_1 m_2 m_3 m_4]}$

Table 1.2: *The spectrum of type IIA and type IIB string theory. Note the position of the Weyl spinor index, α, indicates the chirality.*

Amplitude prescription

A tree level scattering amplitude in the RNS formalism is given by

$$Z = \int \mathcal{D}\varphi \mathcal{O}_1 \cdots \mathcal{O}_N e^{-S_{RNS} - S_{bc} - S_{\beta\gamma} - S_{g.f.g,\chi} - S_{(S)CKV}}, \tag{1.201}$$

where $S_{g.f.g,\chi}$ is the gauge fixing action that sets the metric to the round one and the gravitino to zero. The action $S_{(S)CKV}$ is needed to fix the residual (bosonic and fermionic) gauge invariances. It will be discussed after a short exposition of the

[4]Details about this statement can be found in section 3.2

vertex operators. The case that all external excitations are in the NS sector is dealt with first. As in the bosonic string the vertex operators should be invariant under the symmetries of the action, in particular under the $\mathcal{N}=1$ supersymmetry. The general form of a vertex operator (after fixing the metric and the gravitino), written as an integral over two (bosonic) dimensional superspace, is given by

$$\int d^2z d\theta d\bar{\theta}(V(z)+\theta U(z))(\tilde{V}(\bar{z})+\bar{\theta}\tilde{U}(\bar{z}))e^{ik\cdot X(z,\bar{z})}. \tag{1.202}$$

The two superfield components are related by a supersymmetry transformation

$$U(w)=\int \frac{dz}{2\pi i}G(z)V(w), \tag{1.203}$$

where the supersymmetry current is given by

$$G(z)=G_{mat}(z)+G_{gh}(z), \tag{1.204}$$

The residual gauge invariance includes the super conformal Killing vectors. Recall that the bosonic residual invariance could be eliminated by fixing three (bosonic) coordinates. Similarly the fermionic residual gauge invariance can be removed by fixing two fermionic coordinates. The number two follows from the fact that $P_{\frac{1}{2}}$ has two zero modes on the sphere [3]. Hence the action $S_{(S)CKV}$ is given by

$$S_{(S)CKV}=\sum_{i=1}^{3} B_a^i(\sigma_i^a-\hat{\sigma}_i^a)+\sum_{j=1}^{2} C_s^j\theta_j^s, \tag{1.205}$$

where s is a 2d spinor index. The second term fixes the residual supersymmetry. The measure $\mathcal{D}\varphi$ in (1.201) includes all fields, in particular it contains a factor $d^2\sigma_1\cdots d^2\sigma_N d^2\theta_1\cdots d^2\theta_N$. After the functional integrations, that set the metric and the gravitino to their fixed values, have been performed, the remaining action is a CFT and has a holomorphic and an antiholomorphic sector. After integrating out the auxiliary field C_a^j and the θ_j's, one sees that there are two V's and $N-2$ U's in the amplitude prescription.

The functional integral (1.201) can be processed by integrating out the auxiliary fields. This will impose the gauge conditions and hence will put the action in conformal gauge. Moreover it will introduce ghost insertions multiplying the vertex operators and finally it will remove integrations over (both bosonic and fermionic) worldsheet coordinates. The open string N-NS states tree level function is given by

$$\mathcal{A}^{(N)}=\langle c\mathcal{V}_1^{-1}(y_1)c\mathcal{V}_2^{-1}(y_2)c\mathcal{V}_3^0(y_3)\int dy_4\mathcal{V}_4^0(y_4)\cdots\int dy_N\mathcal{V}_N^0(y_N)\rangle+(\mathcal{V}_1\leftrightarrow\mathcal{V}_2), \tag{1.206}$$

$$\mathcal{V}^{-1}(y)=e^{-\phi}V(y),\quad \mathcal{V}^0(y)=U(y). \tag{1.207}$$

Note the superscripts denote the $\beta\gamma$ charge. A number of comments about this formula are in order:

- The three c insertions come from the Faddeev-Popov action (cf. (1.82)) in exactly the same fashion as in the bosonic string. Recall that this c is actually a $\delta(c)$. In the RNS string there are also two factors of $\delta(\gamma)$, which come from the bosonic discrete ghosts. After bosonization these become $e^{-\phi}$.

- Since both V_{NS} and $c\mathcal{V}^{-1}$ are in BRST cohomology at $k^2 = 0$ they must be equal. Hence

$$V_{NS} = c\mathcal{V}^{-1}. \tag{1.208}$$

- The explicit form of the integrated vertex operators is obtained from (1.203).

- There exists an infinite tower of guises of the vertex operators, each with different $\beta\gamma$ charge, $\cdots, \mathcal{V}^{-1}, \mathcal{V}^{0}, \mathcal{V}^{1}, \cdots$. One can go from one to the next by using the so-called picture changing operator, X [8]:

$$X \cdot \mathcal{V}^{n}(y) = \mathcal{V}^{n+1}(y). \tag{1.209}$$

So in fact the vertex operators in section 1.3.1 do not comprise a complete list. However one can prove that as long as the total $\beta\gamma$ charge is equal to minus two, the amplitude is independent of the pictures of the individual vertex operators. So the omitted vertex operators represent states that were listed in section 1.3.1.

- To be able to compute amplitudes involving Ramond states note that V_R can be written as

$$V_R = c\mathcal{V}^{-\frac{1}{2}}. \tag{1.210}$$

This allows for the construction of an infinite tower vertex operators with half integer picture for the R states via (1.209). An arbitrary amplitude involving both NS and R states is given by

$$\mathcal{A}^{(N)} = \langle c\mathcal{V}_1^{i_1}(y_1) c\mathcal{V}_2^{i_2}(y_2) c\mathcal{V}_3^{i_3}(y_3) \int dy_4 \mathcal{V}_4^{i_4}(y_4) \cdots \int dy_N \mathcal{V}_N^{i_N}(y_N)\rangle + (1 \leftrightarrow 2), \tag{1.211}$$

where the pictures $i_1, \cdots, i_N \in \frac{1}{2}\mathbb{Z}$. When the k-th picture, i_k, is half integer valued the k-th state is a Ramond state. In order to find a nonzero answer one has to ensure that the total picture, i.e. the sum of all i_k's, is equal to -2.

- In [3] it was shown the amplitude prescription is independent of the gauge choice. Note this was not guaranteed due to the ad hoc way of introducing the R state vertex operators.

1.3.2 Higher genus and limitations of RNS

At higher genus the gauge fixing condition for the metric is the same as in the bosonic string. The fixing of the local worldsheet supersymmetry deserves some more explanation. In the same manner as there were two different ways to put the spinor field ψ_m on the infinite leg representing an external state, there are 2^{2g} topologically distinct ways to put a spinor field on a worldsheet of genus g. This number is obtained by noting one has a choice for every non trivial cycle:

$$\chi_\alpha(e^{2\pi i}z) = \pm\chi_\alpha(z). \tag{1.212}$$

Local diffeomorphisms or local worldsheet supersymmetry can never change the spin structure of the gravitino field. Therefore one can only gauge fix a given gravitino to another one with the same spin structure, as depicted in figure 1.11. This results in a sum over spin structures in the amplitude prescription.

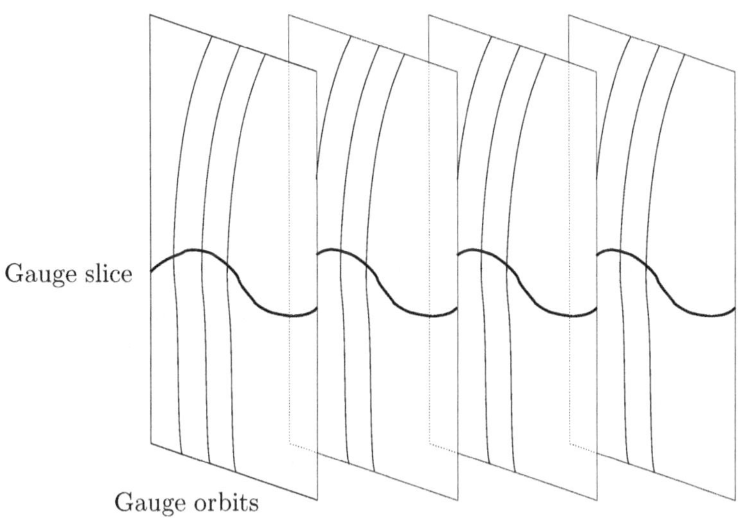

Figure 1.11: *Gravitino space consists of a number of disconnected components. This case would be gravitino space for a genus one surface, a torus.*

Schematically the higher loop amplitude prescription is given by

$$S(1;\cdots;n) = \sum_{\chi,\gamma} e^{\lambda\chi} \int_{\chi,\gamma} d^{n_b}t d^{n_f}\nu \left\langle \prod_{j=1}^{n_b} B_j \prod_{a=1}^{n_f} \delta(B_a) \prod_{i=1}^{n} V_i \right\rangle, \tag{1.213}$$

where the sum is over topologies χ and spin structures γ, n_b is the number of bosonic moduli and n_f is the number of fermionic moduli, B_j is some fermionic object containing the BRST ghosts b and β and the Beltrami differentials. The ghosts b and β are contained in the bosonic object B_a, in addition these depend on the supersymmetric analog of the Beltrami differentials. The precise form of these insertions at two loops and higher is hard to derive. There have been numerous attempts to write them down based on educated guesswork (see introduction of [11] and references therein), but the authors of [3] showed the prescriptions in all of these proposals were not independent of the gauge choice. The same authors derived the insertions from a first principles derivation in a series of papers [12, 13, 14, 15], conveniently summarised in [11]. This shows the importance of a first principles derivation and incidentally provides motivation for the first principles derivation of the pure spinor formalism in chapter 4.

In conclusion the two main problems of the RNS formalism in a flat background are:

1. In order to compute a string diagram of genus g one needs to perform 2^{2g} integrations due to the sum over spin structures. This is a consequence of the presence of worldsheet spinors.

2. The insertions in higher loop diagrams are rather complicated, also in large part due to the presence of worldsheet spinors.

The pure spinor formalism, which will be introduced in the next chapters does not have worldsheet spinors. Hence the above issues do not play a role in the pure spinor formalism. The presence of worldsheet spinors causes one more problem. The total amplitude is the sum over genera, therefore it is important to know the relative factors between the diagrams. The precise value of these factors also plays an important role in checking a conjectured symmetry of string theory, S-duality [10]. It is, however, difficult to compute the overall coefficient of string diagrams. These coefficients include functional determinants (cf. (1.26)). Note that the eigenstates of the kinetic operators, for example ∇^2, depend on the genus of the surface and hence the determinant of ∇^2 also depends on the genus. One also needs to compute functional determinants for the kinetic operators of the worldsheet spinors, which involves bosonisation. For the four point function the overall coefficient has been computed in the RNS formalism in [16]. The two-loop four-point [17] has only been determined up to an overall factor due to unknown factors in the bosonisation formulae of [18]. However in [16] the normalisation of the two-loop amplitude has been obtained by the indirect method of factorisation to lower-loop amplitudes.

There is yet another problem with the RNS formalism, which is related to string theory in a curved background. Within RNS it is hard to generalise (1.146) to backgrounds that involve nonzero RR fields, since the vertex operators for the RR states involve spin half fields. Furthermore the fact that states are represented by an infinite tower of vertex operators related by picture changing complicates the problem. In the pure spinor formalism a generalisation has been written down in [19], which might not come as a surprise since the pure spinor formalism is manifestly spacetime supersymmetric. This thesis deals with string theory in flat backgrounds, therefore this point will not be elaborated on. Nonetheless it is an important motivation to study the pure spinor formalism.

Chapter 2

Pure spinor formalism

This chapter introduces the pure spinor formalism in a flat background. The worldsheet action is an educated guess originally written down by Berkovits. His starting point was not an analog of the Polyakov action, i.e. an action with $2d$ diff $\times$ Weyl invariance, instead he directly wrote down an analog of the worldsheet action in conformal gauge. This means that the action must have a conformal symmetry, zero central charge and a nilpotent fermionic operator that is used to define the spectrum, similar to the way a BRST operator defines a spectrum. Berkovits' proposal satisfies these conditions and on top of that it exhibits manifest spacetime supersymmetry and the worldsheet fields are free. This chapter will discuss the explicit form of the action and some of its properties. Also the prescription for computing scattering amplitudes is provided. This chapter does not contain any explicit computations using this prescription. A good exposition of computations can be found in [20], section 5.1.2 of this thesis also contains some computations.

A number of years after the pure spinor formalism was introduced, Berkovits presented a different but similar formalism. To distinguish the two, the original one was renamed to minimal pure spinor formalism and the modification, the non-minimal pure spinor formalism. The latter was introduced to get rid of some awkward features of the former which will be discussed in due course. Both formalisms are described below. The most recent loop computations, which are also the more complicated ones, have only been performed in the non-minimal formalism, cf. section 2.3 for a precise overview.

This chapter utilises a lot of basic (mathematical) techniques that may or may not be familiar to the reader. In any case these techniques are explained in detail in the next chapter, which can serve either as a necessary addition for a reader new to the subject or as a useful reference for an expert.

2.1 Minimal pure spinor formalism

The worldsheet action in the minimal pure spinor formalism for the left movers in conformal gauge and flat target space is given by

$$S = \int d^2z \left(\frac{1}{2}\partial x^m \bar{\partial} x_m + p_\alpha \bar{\partial}\theta^\alpha - w_\alpha \bar{\partial}\lambda^\alpha \right), \tag{2.1}$$

with $m = 0, \ldots, 9$ and $\alpha = 1, \ldots, 16$. The fields p_α and w_α have conformal weight one and are Weyl spinors, θ^α and λ^α have conformal weight zero and are Weyl spinor of opposite chirality. In addition λ^α is a pure spinor, i.e. it satisfies

$$\lambda^\alpha \gamma^m_{\alpha\beta} \lambda^\beta = 0, \tag{2.2}$$

where $\gamma^m_{\alpha\beta}$ are the ten dimensional Pauli matrices, which are defined in section 3.2. The decomposition of a Weyl spinor under the $SU(5)$ subgroup, $\mathbf{16} \to \mathbf{1} \oplus \bar{\mathbf{10}} \oplus \mathbf{5}$, which is used intensively throughout this work, is also discussed there. Since the worldsheet action consists of two $\beta\gamma$ systems quantisation seems straightforward, but λ^α is a pure spinor and therefore the λw part is actually a curved $\beta\gamma$ system [21]. To deal with this we work on a patch in pure spinor space that is defined by $\lambda^+ \neq 0$. On this patch the pure spinor condition expresses λ^a in terms of λ_{ab} and λ^+, with $a, b = 1, \ldots, 5$. The solution is (in $SU(5)$ covariant components)

$$\lambda^a = \frac{1}{8}\frac{1}{\lambda^+}\epsilon^{abcde}\lambda_{bc}\lambda_{de}. \tag{2.3}$$

A constraint on fields in the action induces a gauge invariance on the conjugate fields. In this case the gauge transformations are given by

$$\delta w_\alpha = \Lambda_m \gamma^m_{\alpha\beta}\lambda^\beta. \tag{2.4}$$

In the original papers, e.g. [22], this gauge invariance is dealt with by only using gauge invariant quantities. This means w_α can only appear in the Lorentz current N^{mn}, the ghost number current J and the stress energy tensor $T_{(\lambda w)}$:

$$N^{mn} = \frac{1}{2} w_\alpha (\gamma^{mn})^\alpha{}_\beta \lambda^\beta, \quad J = w_\alpha \lambda^\alpha, \quad T_{(\lambda w)} = w_\alpha \partial\lambda^\alpha. \tag{2.5}$$

Since the λw part of the action is not free due to the pure spinor constraint it is not obvious what the OPE between w and λ will be. One way to proceed is by properly fixing the gauge invariance of (2.4). By making the gauge choice $w_a = 0$ and employing BRST methods, one can replace $\int d^2z w_\alpha \bar{\partial}\lambda^\alpha$ by the free action,

$$\int d^2z (\omega_+ \bar{\partial}\lambda^+ + \frac{1}{2}\omega^{ab}\bar{\partial}\lambda_{ab}). \tag{2.6}$$

The details can be found in section 3.3.2. One might have expected BRST ghosts associated to the gauge fixing of w_α. It turns out these can be integrated out. As a check of the validity of this procedure the OPE of the Lorentz currents $(N^{mn}|_{w_\alpha=0})$ should give rise to the Lorentz algebra. Using (2.6) one finds

$$N^{mn}(z)\lambda^\alpha(w) \sim \frac{1}{z-w}\frac{1}{2}(\gamma^{mn}\lambda)^\alpha, \quad J(z)\lambda^\alpha(w) \sim \frac{1}{z-w}\lambda^\alpha, \tag{2.7}$$

$$N^{mn}(z)N^{pq}(w) \sim \frac{-3}{(z-w)^2}(\eta^{n[p}\eta^{q]m}) + \frac{1}{z-w}(\eta^{n[p}N^{q]m} - \eta^{m[p}N^{q]n}),$$

$$J(z)J(w) \sim \frac{-4}{(z-w)^2}, \quad J(z)N^{mn}(w) \sim \text{regular},$$

$$N^{mn}(z)T(w) \sim \frac{1}{(z-w)^2}N^{mn}(w), \quad J(z)T(w) \sim \frac{-8}{(z-w)^3} + \frac{1}{(z-w)^2}J(w).$$

The explicit computations can be found in appendix 3.3 and it should be noted there are subtleties regarding the double poles in the OPE. There is freedom to add conserved currents to the Lorentz currents without changing the single poles, which must have the above form for Lorentz currents. However if one demands that N^{mn} is a primary field with Lorentz level -3, the above OPE's follow unambiguously. One wants the level of the Lorentz current to be -3, since this implies that the level of the total non X sector is $4-3=1$ which coincides with the level of the RNS $\psi\psi$ Lorentz current. The factor of -8 of the triple pole in the JT OPE implies at tree level only correlators with total J charge -8 will be nonzero (cf. (1.181)). The OPE's for the matter variables can be straightforwardly derived from (2.1):

$$x^m(z)x^n(w) \sim -\eta^{mn}\log|z-w|^2, \quad p_\alpha(z)\theta^\beta(w) \sim \delta^\alpha_\beta\frac{1}{z-w}. \tag{2.8}$$

The action (2.1) is invariant under a nilpotent fermionic symmetry generated by[1]

$$Q_S = \oint dz\lambda^\alpha d_\alpha, \tag{2.9}$$

where

$$d_\alpha = p_\alpha - \frac{1}{2}\gamma^m_{\alpha\beta}\theta^\beta\partial x_m - \frac{1}{8}\gamma^m_{\alpha\beta}\gamma_{m\ \gamma\delta}\theta^\beta\theta^\gamma\partial\theta^\delta. \tag{2.10}$$

The transformations it generates are given by

$$\delta x^m = \lambda\gamma^m\theta, \quad \delta\theta^\alpha = \lambda^\alpha, \quad \delta\lambda^\alpha = 0, \quad \delta d_\alpha = -\Pi^m(\gamma_m\lambda)_\alpha, \quad \delta w_\alpha = d_\alpha, \tag{2.11}$$

where $\Pi^m = \partial x^m + \frac{1}{2}\theta\gamma^m\partial\theta$ is the supersymmetric momentum and again we restrict to the left movers (so in particular, the full transformation for x^m contains a similar additive term with right moving fields).

It seems very natural to consider Q_S as a BRST operator that showed up after fixing a worldsheet symmetry, in particular diffeomorphism invariance. However to

[1]The unconventional subscript S is used to distinguish this operator from another nilpotent fermionic operator which will appear in chapter 4.

date nobody has succeeded in substantiating this conjecture, although the authors of [23] describe how it is possible to obtain the pure spinor formalism as a twisted version of a gauge fixed string theory with diffeomorphism invariance. In chapter 4 the worldsheet action in conformal gauge will be derived by gauge fixing a worldsheet action with diffeomorphism symmetry. However the $2d$ coordinate invariant action is already invariant under Q_S, the gauge fixing of the diffeomorphisms gives rise to a second nilpotent fermionic operator. This is a different point of view where Q_S is not a BRST operator of fixing $2d$ coordinate invariance.

The main motivation to introduce the pure spinor formalism is its manifest supersymmetry. This symmetry is generated by

$$q_\alpha = \oint dz(p_\alpha + \frac{1}{2}\gamma^m_{\alpha\beta}\theta^\beta \partial x_m + \frac{1}{24}\gamma^m_{\alpha\beta}(\gamma_m)_{\gamma\delta}\theta^\beta\theta^\gamma\theta^\delta). \tag{2.12}$$

2.1.1 Spectrum

Physical states are defined as element of the cohomology of Q_S with $J_{\lambda w}$ charge one and conformal weight zero. In theories derived from a worldsheet diffeomorphism invariant action, the conformal weight constraint follows from the condition that physical states must be annihilated by the BRST operator. In the case of the pure spinor action the operator Q_S does not impose a constraint on the conformal weight and it has to be included by hand. In chapter 4 the origin of conformal weight constraint is explained from first principles in the case of the pure spinor formalism. The reason to look at ghost number one states is more subtle. At least one can say that the cohomology at this $J_{\lambda w}$ charge yields the super-Maxwell multiplet (for the open string).

Hence elements of the physical spectrum satisfy:

$$Q_S V(z) = 0, \quad V(z) \sim V(z) + Q_S \Omega(z). \tag{2.13}$$

Let us focus on the massless spectrum. The most general vertex operator (before imposing the above conditions) at $J_{\lambda w}$ charge one with conformal dimension zero and $k^2 = 0$ is given by

$$V(z) = e^{ik\cdot X(z,\bar{z})}\lambda^\alpha(z) A_\alpha(\theta(z)). \tag{2.14}$$

A number of comments are in order

- For the X sector one uses the standard operators (1.116) and note that the weight is only non positive when no derivatives on X are present.

- The weight of the p, θ and w, λ sector is only non positive when V only contains λ and θ.

- The total weight of V in the massless case can only be zero if it only consists of weight zero fields. This determines the form of (2.14) completely.
- Since there are no negative weight fields, there is no tachyon present in the spectrum. There is an infinite tower of massive states, but these will not be considered in this thesis.

After using the gauge invariance to set a number of components to zero the solution to (2.13) is given by [24]

$$V = \lambda^\alpha A_\alpha(x, \theta), \tag{2.15}$$

where

$$A_\alpha(x, \theta) = e^{ik\cdot x}(\frac{1}{2}a_m(\gamma^m\theta)_\alpha - \frac{1}{3}(\xi\gamma_m\theta)(\gamma^m\theta)_\alpha + \cdots), \tag{2.16}$$

where a_m and ξ^α are the polarisations and k^m is the momentum. They satisfy $k^2 = k^m a_m = k^m(\gamma_m\xi)_\alpha = 0$, there is a residual gauge invariance $a_m \to a_m + k_m\omega$ and the ellipsis contains products of k^m with a_m or ξ^α. The operator $V(z)$ can be used as unintegrated vertex operator.

The integrated vertex operators can again be obtained by an educated guess based on comparison with the bosonic string and/or the RNS string. In those theories the integrated vertex operator satisfies

$$Q_S U(z) = \partial V(z) \tag{2.17}$$

This equation also has a solution in the pure spinor formalism, which is given by

$$U = \partial\theta^\alpha A_\alpha(x, \theta) + \Pi^m A_m(x, \theta) + d_\alpha W^\alpha(x, \theta) + \frac{1}{2}N^{mn}\mathcal{F}_{mn}(x, \theta), \tag{2.18}$$

with

$$\begin{aligned} A_m &= \frac{1}{8}D_\alpha\gamma_m^{\alpha\beta}A_\beta, & (2.19)\\ W^\beta &= \frac{1}{10}\gamma_m^{\alpha\beta}(D_\alpha A^m - \partial^m A_\alpha), & (2.20)\\ \mathcal{F}_{mn} &= \frac{1}{8}D_\alpha(\gamma_{mn})^\alpha{}_\beta W^\beta, & (2.21) \end{aligned}$$

where $D_\alpha = \frac{\partial}{\partial\theta^\alpha} + \frac{1}{2}\theta^\beta\gamma^m_{\alpha\beta}\partial_m$.

2.1.2 Tree-level prescription

Originally the amplitude prescription in the pure spinor formalism was motivated by analogy to the bosonic string. The guiding principles are given by

- There are three unintegrated vertex operators and $N - 3$ integrated ones to deal with the CKG.

- The total $J_{\lambda w}$ charge must equal the charge anomaly (2.7).

The N point open string tree-level amplitude prescription presented in [22] satisfies the above guiding principles.

$$\mathcal{A} = \langle V_1(z_1)V_2(z_2)V_3(z_3)\int dz_4U_4(z_4)\cdots\int dz_NU_N(z_N)Y_{C_1}(y_1)\cdots Y_{C_{11}}(y_{11})\rangle =$$

$$\int[\mathcal{D}^{10}x][\mathcal{D}^{16}d][\mathcal{D}^{16}\theta][\mathcal{D}^{11}\lambda][\mathcal{D}^{11}w]V_1(z_1)V_2(z_2)V_3(z_3)\int dz_4U_4(z_4)\cdots\int dz_NU(z_N)$$
$$Y_{C_1}(y_1)\cdots Y_{C_{11}}(y_{11})e^{-S}, \tag{2.22}$$

where $[\mathcal{D}\phi]$ denotes functional integration over the field ϕ. The functional integration over x^m have been studied in detail and the same correlation functions appear in the RNS formalism. This factor will be ignored when it is not relevant to the computation.

Y_C are the picture changing operators (PCOs):

$$Y_C(y) = C_\alpha\theta^\alpha(y)\delta(C_\beta\lambda^\beta(y)), \tag{2.23}$$

where C_α is a constant spinor. The presence of the PCOs in the amplitude prescription is explained from first principles in chapter 4. In short, they come from fixing a gauge invariance due to the zero modes of the weight zero fields, $\lambda^\alpha, \theta^\alpha$. Note the weight one fields do not have zero modes[2] at tree level. At higher loops there will also be PCOs for these fields. Since the PCOs are introduced as a gauge fixing term, amplitudes should be independent of the constant tensors C_α. The name picture changing operator was also given to an operator in the RNS formalism (1.209). These operators change the (bosonic) ghost number of the vertex operators. The λ, w sector can be seen as a ghost sector since it is not part of the ten dimensional superspace and they have the "wrong" spin-statistics relation. Since Y_C change the $J_{\lambda w}$ charge by one, these operators were also named picture changing operators.

The functional integral (2.22) is evaluated by first using the OPE's of (2.7) and (2.8). Note that this operation reduces the total conformal dimension of the worldsheet fields involved in the OPE. For example in the p, θ OPE, the conformal weight of $p_\beta(z)\theta^\alpha(w)$ is one and the conformal weight of δ^α_β is zero. Thus in the end the correlator only contains worldsheet fields of weight zero. This can be evaluated by replacing the fields by their zero modes and performing the zero mode integrations. The justification for this step is given in section 4.6.

After integrating out the nonzero modes the amplitude reduces to

$$\mathcal{A} = \int[d\lambda]d^{16}\theta\lambda^\alpha\lambda^\beta\lambda^\gamma f_{\alpha\beta\gamma}(\theta)(C^1\theta)\delta(C^1\lambda)\cdots(C^{11}\theta)\delta(C^{11}\lambda), \tag{2.24}$$

[2]This can be inferred from the Riemann Roch theorem (1.74) and the fact that weight zero fields have precisely one zero mode at any genus.

where $f_{\alpha\beta\gamma}$ depends on all the polarisations and momenta. Note the functional integration of x^m is omitted here as will be done in all computations in this thesis. A priori $f_{\alpha\beta\gamma}$ also depends on z_1, z_2, z_3. Of course we expect the final result to be independent of these coordinates. Also note all the fields are zero modes including those in the measure. $[d\lambda]$ is the unique Lorentz invariant measure of +8 ghost number on the space of pure spinors (cf. section 3.4). It is given by

$$[d\lambda]\lambda^\alpha\lambda^\beta\lambda^\gamma = d\lambda^{\alpha_1} \wedge \cdots \wedge d\lambda^{\alpha_{11}} (\epsilon T)^{\alpha\beta\gamma}_{\alpha_1\cdots\alpha_{11}}, \tag{2.25}$$

where

$$(\epsilon T)^{\alpha\beta\gamma}_{\alpha_1\cdots\alpha_{11}} = \epsilon_{\alpha_1\cdots\alpha_{16}} \gamma_m^{\alpha\alpha_{12}} \gamma_n^{\beta\alpha_{13}} \gamma_p^{\gamma\alpha_{14}} (\gamma^{mnp})^{\alpha_{15}\alpha_{16}}. \tag{2.26}$$

Note no gamma trace is subtracted. This tensor is already gamma matrix traceless as explained in section 3.4.

Lorentz invariance

The PCOs contain constant spinors. Therefore the prescription is not manifestly Lorentz invariant and one has to check Lorentz invariance by hand. The Lorentz variation of one PCO is given by:

$$M^{mn}Y_C = \frac{1}{2}(C\gamma^{mn}\theta)\delta(C\lambda) + \frac{1}{2}(C\theta)(C\gamma^{mn}\lambda)\delta'(C\lambda) = Q_S[\frac{1}{2}(C\gamma^{mn}\theta)(C\theta)\partial\delta(C\lambda)]. \tag{2.27}$$

The last equality shows that the Lorentz variation of the PCO is Q_S exact. This decouples if all other insertions are Q_S closed and $\langle Q_S K\rangle$ vanishes for all K. The second condition is satisfied because after integrating out the non-zero modes $\langle Q_S X\rangle$ reduces to

$$\int [d\lambda] d^{16}\theta \lambda^\alpha\lambda^\beta\lambda^\gamma D_\alpha f_{\beta\gamma}(\theta) C^1\theta\delta(C^1\lambda)\cdots C^{11}\theta\delta(C^{11}\lambda) = 0, \tag{2.28}$$

because $\int d^{16}\theta D_\alpha g(\theta) = 0$ for any function g. The first condition is more subtle. The vertex operators are Q_S closed, due to the physical state condition (2.13). In order to see whether the PCOs are closed consider

$$Q_S Y_C = C_\alpha\lambda^\alpha\delta(C_\beta\lambda^\beta). \tag{2.29}$$

This seems to be zero, but there are subtleties due to the presence of factors of λ in the denominator form the measure (2.25). A detailed exposition of these subtleties can be found in chapter 5.

It is possible to restore manifest Lorentz invariance by integrating over all possible choices for C. This guarantees the prescription is Lorentz invariant. However it does not guarantee that Q_S exact states will decouple. After including the C integral (2.24) becomes

$$\mathcal{A} = \int [dC][d\lambda] d^{16}\theta \lambda^\alpha\lambda^\beta\lambda^\gamma f_{\alpha\beta\gamma}(\theta)(C^1\theta)\delta(C^1\lambda)\cdots(C^{11}\theta)\delta(C^{11}\lambda). \tag{2.30}$$

2.1.3 One and higher-loop prescription

Let us start with giving the one-loop amplitude prescription. Compared to a tree-level function a one-loop function exhibits three new features:

- PCOs for the weight one worldsheet fields p, w,
- zero mode integrals over p, w,
- a composite b ghost constructed out of the worldsheet fields from (2.1).

The first two points are direct consequences of the presence of a zero mode of weight one fields on the torus. The new PCOs are given in terms of the gauge invariant quantities N^{mn} and J:

$$Z_B(z) = \frac{1}{2}B_{mn}\lambda(z)\gamma^{mn}d(z)\delta(B_{mn}N^{mn}(z)), \quad Z_J(z) = \lambda^\alpha(z)d_\alpha(z)\delta(J(z)). \tag{2.31}$$

Note that the picture raising[3] operators, Z_B and Z_J, are Q_S-closed without subtleties:

$$Q_S Z_B = \frac{1}{4}B_{mn}\lambda\gamma^{mn}dB_{m'n'}\lambda\gamma^{m'n'}d\delta'(B_{pq}N^{pq}) = \frac{1}{4}(B_{mn}\lambda\gamma^{mn}d)^2\delta'(B_{pq}N^{pq}) = 0. \tag{2.32}$$

This vanishes because it contains the square of a fermionic quantity. Let us also record the Lorentz variation of Z_B,

$$M^{mn}Z_B = Q_S[2\eta^{p[m}\delta^{n]}_r B_{pq}N^{qr}\delta(BN)], \tag{2.33}$$

which is Q_S exact.

All string theory amplitude prescriptions at one loop contain a b ghost which satisfies

$$\{Q_S, b(z)\} = T(z). \tag{2.34}$$

In the RNS formalism this field appears as one of the two reparametrisation Faddeev Popov ghosts and note that at one loop there should be one (holomorphic) b ghost insertion to absorb the zero mode (cf. table 1.1). In the bosonic string amplitude prescription, which the pure spinor amplitude prescription is analogous to, this b insertion enters through $(b, \partial_\tau g)$, where the brackets have been defined in (1.69) and τ is the modulus of a genus one surface. While the full derivation of the form of this insertion will be given in chapter 4, it is possible to show this insertion in consistent with BRST invariance, since its variation equals a total derivative in moduli space which vanishes upon integrating over the moduli:

$$Q_S(b, \partial_\tau g)e^{-S} = (T, \partial_\tau g)e^{-S} = \frac{\partial S}{\partial g_{ab}}\frac{\partial g_{ab}}{\partial \tau}e^{-S} = -\frac{\partial}{\partial \tau}e^{-S}. \tag{2.35}$$

[3]These are called raising operators because they have $+1$ $J_{\lambda w}$ charge, hence they raise the picture in the language of (1.209).

The derivative of the metric with respect to the modulus is called a Beltrami differential, μ, and on higher genus surfaces the Beltrami differential has an index that runs over the number of moduli, μ_k.

In the pure spinor formalism, however, the b ghost is constructed out of the worldsheet fields from (2.1) as explained from first principles in chapter 4. It is not possible to solve equation (2.34) in the minimal pure spinor formalism, because of ghost number (J charge) conservation combined with gauge invariance of objects containing w_α. The former implies b must have ghost number minus one and since there are no gauge invariant quantities with negative ghost number the latter rules out any solution. A resolution to this problem is combining the (composite) b field with a PCO, Z_B, such that

$$\{Q_S, \tilde{b}_B(u,z)\} = T(u)Z_B(z). \tag{2.36}$$

This equation ensures the Q_S variation of the b ghost vanishes after integrating over moduli space. The solution is given by

$$\tilde{b}_B(u,z) = b_B(u) + T(u)\int_u^z dv B_{pq}\partial N^{pq}(v)\delta(BN(v)). \tag{2.37}$$

The local b ghost, $b_B(u)$, is a composite operator, constructed out of the worldsheet fields:

$$b_B(z) = b_{B0}(z)\delta(BN(z)) + b_{B1}(z)\delta'(BN(z)) + b_{B2}(z)\delta''(BN(z)) + b_{B3}(z)\delta'''(BN(z)), \tag{2.38}$$

where the primes denote derivatives, $BN \equiv B_{mn}N^{mn}$ and

$$\begin{aligned}
b_{B0} &= \frac{1}{2}G\gamma^{mn}dB_{mn} - \frac{1}{2}H^{\alpha\beta}(\gamma^p\gamma^{mn})_{\alpha\beta}\Pi_p B_{mn} + \qquad (2.39)\\
&\quad \frac{1}{2}K^{\alpha\beta\gamma}(\gamma^p\gamma^{mn})_{\beta\gamma}(\gamma_p\partial\theta)_\alpha B_{mn} + \frac{1}{2}S^{\alpha\beta\gamma}(\gamma^p\gamma^{mn})_{\beta\gamma}(\gamma^p\partial\lambda)_\alpha B_{mn},\\
b_{B1} &= \frac{1}{4}H^{\alpha\beta}(Bd)_\alpha(Bd)_\beta + \qquad (2.40)\\
&\quad \frac{1}{4}K^{\alpha\beta\gamma}(\gamma^p\gamma^{mn})_{\beta\gamma}(Bd)_\alpha\Pi_p B_{mn} + \frac{1}{4}K^{\alpha\beta\gamma}(\gamma^p\gamma^{mn})_{\alpha[\beta}(Bd)_{\gamma]}\Pi_p B_{mn} +\\
&\quad \frac{1}{4}L^{\alpha\beta\gamma\delta}[((\gamma^p\gamma^{mn})_{\gamma\delta}(Bd)_{[\alpha}(\gamma_p\partial\theta)_{\beta]} - (\gamma^p\gamma^{mn})_{\beta[\gamma}(Bd)_{\delta]}(\gamma_p\partial\theta)_\alpha)B_{mn} -\\
&\quad ((\gamma^s\gamma^{rq})_{\alpha[\beta}(\gamma^p\gamma^{mn})_{\gamma]\delta} + (\gamma^s\gamma^{rq})_{\alpha\delta}(\gamma^p\gamma^{mn})_{\beta\gamma})\Pi_p B_{mn}\Pi_s B_{qr}],\\
b_{B2} &= -\frac{1}{8}K^{\alpha\beta\gamma}(Bd)_\alpha(Bd)_\beta(Bd)_\gamma - \frac{1}{8}L^{\alpha\beta\gamma\delta}((\gamma^p\gamma^{mn})_{\gamma\delta}(Bd)_\beta(Bd)_\alpha + \qquad (2.41)\\
&\quad (\gamma^p\gamma^{mn})_{\beta[\gamma}(Bd)_{\delta]}(Bd)_\alpha + \frac{1}{2}(\gamma^p\gamma^{mn})_{\alpha[\delta}(Bd)_\gamma(Bd)_{\beta]})\Pi_p B_{mn},\\
b_{B3} &= -\frac{1}{16}L^{\alpha\beta\gamma\delta}(Bd)_\alpha(Bd)_\beta(Bd)_\gamma(Bd)_\delta, \qquad (2.42)
\end{aligned}$$

where $(Bd)_\alpha \equiv B_{mn}(\gamma^{mn}d)_\alpha$ and G, H, K, L are given in appendix 3.6.

The one-loop amplitude prescription in the minimal pure spinor formalism is given by

$$\mathcal{A}^{(N)} = \int d^2\tau \langle | \int d^2u \mu(u) \tilde{b}_{B^1}(u, z_1) \prod_{P=2}^{10} Z_{B^P}(z_P) Z_J(z_{11}) \prod_{I=1}^{11} Y_{C_I}(y)|^2 \tag{2.43}$$

$$V_1(t_1) \prod_{T=2}^{N} \int d^2t_T U_T(t_T)\rangle,$$

where $\mu(u)$ is the zz component of the Beltrami differential.

Above one loop there are no conformal killing vectors anymore, so that there is no unintegrated vertex operator. The number of metric moduli at genus g is given by $6g-6$ (cf. table 1.1) and all conformal weight one fields have g zero modes each. This leads to the multiloop amplitude prescription of the minimal pure spinor formalism:

$$\mathcal{A}^{(N)} = \int d^2\tau_1 \cdots d^2\tau_{3g-3} \langle | \prod_{P=1}^{3g-3} \int d^2u_P \mu_P(u_P) \tilde{b}_{B_P}(u_P, z_P) \tag{2.44}$$

$$\prod_{P=3g-2}^{10g} Z_{B_P}(z_P) \prod_{R=1}^{g} Z_J(v_R) \prod_{I=1}^{11} Y_{C_I}(y_I)|^2 \prod_{T=1}^{N} \int d^2t_T U_T(t_T)\rangle.$$

As described in [22], the amplitudes (2.43) and (2.44) are evaluated by first using the OPE's to remove all fields of nonzero weight. After this step all fields have weight zero. This can be evaluated by replacing the fields by their zero modes and performing the zero mode integrations. Therefore one needs to know how to integrate over the zero modes. For the d, θ, x variables this is standard, so only the integration over λ, N, B, C is discussed.

A typical integral one encounters is given by [22]:

$$\mathcal{A} = \int [d\lambda][dB][dC] \prod_{R=1}^{g} [dN_R] f(\lambda, N_R, J_R, C, B), \tag{2.45}$$

where $[dN]$ is the zero mode measure for (each zero mode of) N_{mn} (cf. section 3.4.2). It must have $J_{\lambda w}$ charge -8, since the JT OPE in (2.7) implies only correlators with a total $J_{\lambda w}$ charge of $8-8g$ can be non vanishing at g loops [8]. It is given by

$$[dN]\lambda^{\alpha_1} \cdots \lambda^{\alpha_8} = dN^{m_1 n_1} \wedge \cdots \wedge N^{m_{10} n_{10}} \wedge dJ R^{\alpha_1 \cdots \alpha_8}_{m_1 n_1 \cdots m_{10} n_{10}}, \tag{2.46}$$

with

$$R^{\alpha_1 \cdots \alpha_8}_{m_1 n_1 \cdots m_{10} n_{10}} \equiv \tag{2.47}$$

$$\gamma^{((\alpha_1 \alpha_2}_{m_1 n_1 m_2 m_3 m_4} \gamma^{\alpha_3 \alpha_4}_{m_5 n_5 n_2 m_6 m_7} \gamma^{\alpha_5 \alpha_6}_{m_8 n_8 n_3 n_6 m_9} \gamma^{\alpha_7 \alpha_8))}_{m_{10} n_{10} n_4 n_7 n_9} + \text{permutations}.$$

The permutations make R antisymmetric under exchange in both $m_i \leftrightarrow n_i$ and $m_i n_i \leftrightarrow m_j n_j$ and the double brackets denote subtraction of the gamma trace. The zero mode integral (2.45) is only nonzero if the function f depends on (λ, N, J, C, B) as

$$f(\lambda, N, J, C, B) = \tag{2.48}$$

$$h(\lambda, N, J, C, B) \prod_{R=1}^{g} \partial^{M_R}\delta(J) \prod_{P=1}^{10}\prod_{R=1}^{g} \partial^{L_{P,R}}\delta(B^P N_R) \prod_{I=1}^{11} \partial^{K_I}\delta(C^I\lambda),$$

where the polynomial h assumes the form

$$(\lambda)^{8g-8+\sum_{I=1}^{11}(K_I+1)} \prod_{R=1}^{g} (J_R)^{M_R} (N_R)^{\sum_{P=1}^{10} L_{P,R}} \prod_{P=1}^{10} (B^P)^{L_{P,R}+1} \prod_{I=1}^{11} (C^I)^{K_I+1}. \tag{2.49}$$

The integration over the zero modes of the pure spinor variables and the constant tensors is defined in [22] as

$$\mathcal{A}^{(N)} = c \prod_{I=1}^{11} \left(\frac{\partial}{\partial\lambda^\delta}\frac{\partial}{\partial C^I_\delta}\right)^{K_I} \prod_{P=1}^{10}\prod_{R=1}^{g} \left(\frac{\partial}{\partial B^P_{pq}}\frac{\partial}{\partial N^{pq}_R}\right)^{L_{P,R}} \prod_{R=1}^{g} \left(\frac{\partial}{\partial J_R}\right)^{M_R} \tag{2.50}$$

$$\frac{\partial}{\partial C^1_{\beta_1}} \cdots \frac{\partial}{\partial C^{11}_{\beta_{11}}} \frac{\partial}{\partial\lambda^{\alpha_1}} \cdots \frac{\partial}{\partial\lambda^{\alpha_3}} (\epsilon T)^{\alpha_1\cdots\alpha_3}_{\beta_1\cdots\beta_{11}}$$

$$\left[R^{\alpha_4\cdots\alpha_{11}}_{m_1n_1\cdots m_{10}n_{10}} \frac{\partial}{\partial\lambda^{\alpha_4}} \cdots \frac{\partial}{\partial\lambda^{\alpha_{11}}} \frac{\partial}{\partial B^1_{m_1n_1}} \cdots \frac{\partial}{\partial B^{10}_{m_{10}n_{10}}} \right]^g h(\lambda, N_R, J_R, C, B),$$

for some proportionality constant c.

2.1.4 Decoupling of Q_S exact states and PCO positions

The amplitude prescriptions, (2.22) and (2.43), put the PCOs at arbitrary points on the worldsheet. Of course the final result cannot depend on these positions, since they do not contain any physical significance. To study the dependence on the insertions point one looks at the worldsheet derivatives of the PCOs:

$$\partial Y_C(y) = Q_S[(C\partial\theta(y))(C\theta(y))\delta'(C\lambda(y))], \tag{2.51}$$

$$\partial Z_B(z) = Q_S[-B_{pq}\partial N^{pq}(z)\delta(BN(z))], \quad \partial Z_J(z) = Q_S[-\partial J(z)\delta(J(z))]. \tag{2.52}$$

These are Q_S exact, like the Lorentz variation of the PCOs. Hence the amplitude is only guaranteed to be independent of these insertions points if Q_S exact states decouple. Due to the subtleties with Q_S closedness of Y_C this is non-trivial. In chapter 5 the problem is completely solved and a proof of decoupling of Q_S exact states is given, hence also proving Lorentz invariance and independence of PCO positions.

2.2 Non-minimal pure spinor formalism

The minimal pure spinor formalism has the desired property of manifest spacetime supersymmetry. However, manifest Lorentz invariance is not present, due to the appearance of the constant spinors/tensors C and B. Furthermore the b ghost equation (2.34) could not be solved. These two problems are resolved in the non-minimal pure spinor formalism.

The non-minimal version of the formalism [25] (see [26] for a review) amounts to introducing a set of non-minimal variables, the complex conjugate $\bar{\lambda}_\alpha$ of λ^α, a fermionic constrained spinor r_β satisfying

$$\bar{\lambda}_\alpha \gamma_m^{\alpha\beta} \bar{\lambda}_\beta = 0, \qquad \bar{\lambda}_\alpha \gamma_m^{\alpha\beta} r_\beta = 0 \tag{2.53}$$

and their conjugate momenta, $\bar{w}^\alpha$ and s^a. Analogous to the minimal formalism these conditions induce a gauge invariance:

$$\delta\bar{w}^\alpha = \bar{\Lambda}^m(\gamma_m\bar{\lambda})^\alpha - \phi^m(\gamma_m r)^\alpha, \quad \delta s^\alpha = \phi^m(\gamma_m\bar{\lambda})^\alpha. \tag{2.54}$$

This implies $\bar{w}^\alpha$ and s^α can only appear in the gauge invariant quantities

$$\bar{N}^{mn} = \frac{1}{2}(\bar{\lambda}\gamma^{mn}\bar{w} - s\gamma_{mn}r), \quad \bar{J} = \bar{\lambda}\bar{w} - sr, \quad T_{\bar{\lambda}\bar{w}} = \bar{w}^\alpha\partial\bar{\lambda}_\alpha - s^\alpha\partial r_\alpha, \tag{2.55}$$

$$S_{mn} = \frac{1}{2}s\gamma_{mn}\bar{\lambda}, \quad S = s\bar{\lambda}.$$

The action (2.1) is modified by the addition of the term S_{nm}:

$$S \to S + S_{nm}, \qquad S_{nm} = \int d^2z \left(-\bar{w}^\alpha\bar{\partial}\bar{\lambda}_\alpha + s^\alpha\bar{\partial}r_\alpha\right) \tag{2.56}$$

and the generator Q_S by

$$Q_S \to Q_S + \oint dz\bar{w}^\alpha r_\alpha. \tag{2.57}$$

This acts on the non-minimal variables as follows

$$\delta\bar{\lambda}_\alpha = r_\alpha, \qquad \delta r_\alpha = 0, \qquad \delta s^\alpha = \bar{w}^\alpha, \qquad \delta\bar{w}^\alpha = 0. \tag{2.58}$$

These transformation rules imply that the cohomology is independent of the non-minimal variables. In other words the vertex operators can always be chosen such that they do not include these variables. A more natural point of view, which will be adopted in chapter 4, is to consider the non-minimal variables as fields that appear in the BRST treatment of gauge freedom due to shifts of the zero modes of the worldsheet fields. This also explains why vertex operators do not depend on the non-minimal fields and why only the zero modes of these fields appear in the path integral. Furthermore the OPE's given in section 2.1 still comprise a complete list, since the new fields do not have non zero modes. The tree-level amplitude prescription is given by

$$\mathcal{A} = \langle V_1(z_1)V_2(z_2)V_3(z_3)\prod_{i=4}^{N}\int dz_i U_i(z_i) e^{-(\lambda(y)\bar{\lambda}(y)+r(y)\theta(y))}\rangle. \tag{2.59}$$

Compared to the minimal case the PCOs have been replaced by

$$\mathcal{N}(y) \equiv e^{(\{Q_S,-\bar{\lambda}(y)\theta(y)\})} = e^{-(\lambda(y)\bar{\lambda}(y)+r(y)\theta(y))}. \tag{2.60}$$

Originally this factor was postulated by Berkovits, but it can also be derived from first principles. This will be done in chapter 4. Unlike the PCOs $\mathcal{N}$ is Q_S closed without subtleties:

$$Q_S e^{-(\lambda\bar{\lambda}+r\theta)} = Q_S[-(\lambda\bar{\lambda}+r\theta)]e^{-(\lambda\bar{\lambda}+r\theta)} = -(\lambda r - r\lambda)e^{-(\lambda\bar{\lambda}+r\theta)} = 0. \tag{2.61}$$

Furthermore amplitude will not depend on the insertion point y since y only appears in a Q_S exact term. More precisely $\mathcal{N}$ can be written as $1 + Q_S\Omega$ for some Ω and all y dependence is in that Ω.

After performing the OPE's between the vertex operators, which results in exactly the same function $f_{\alpha\beta\gamma}$ as the minimal manipulations, all fields can be replaced by their zero modes:

$$\mathcal{A} = \int d^{16}\theta f_{\alpha\beta\gamma}(\theta)\int [d\lambda][d\bar{\lambda}][dr]\lambda^\alpha\lambda^\beta\lambda^\gamma e^{-(\lambda\bar{\lambda}+r\theta)}, \tag{2.62}$$

where $[d\bar{\lambda}]$ and $[dr]$ are Lorentz invariant measures:

$$[d\bar{\lambda}]\bar{\lambda}_\alpha\bar{\lambda}_\beta\bar{\lambda}_\gamma = (\epsilon T)^{\alpha_1\cdots\alpha_{11}}_{\alpha\beta\gamma}d\bar{\lambda}_{\alpha_1}\cdots d\bar{\lambda}_{\alpha_{11}} \tag{2.63}$$

and

$$[dr] = (\epsilon T)^{\alpha\beta\gamma}_{\alpha_1\cdots\alpha_{11}}\bar{\lambda}_\alpha\bar{\lambda}_\beta\bar{\lambda}_\gamma\frac{\partial}{\partial r_{\alpha_1}}\cdots\frac{\partial}{\partial r_{\alpha_{11}}}. \tag{2.64}$$

The invariant tensor (ϵT) with indices in the opposite positions compared to (2.26) is defined by

$$(\epsilon T)^{\alpha_1\cdots\alpha_{11}}_{\alpha\beta\gamma} = \epsilon^{\alpha_1\cdots\alpha_{16}}\gamma^m_{\alpha\alpha_{12}}\gamma^n_{\beta\alpha_{13}}\gamma^p_{\gamma\alpha_{14}}(\gamma_{mnp})_{\alpha_{15}\alpha_{16}}. \tag{2.65}$$

We know $\int[d\lambda][d\bar{\lambda}][dr]\lambda^\alpha\lambda^\beta\lambda^\gamma e^{-(\lambda\bar{\lambda}+r\theta)}$ must be a Lorentz tensor with three spinor indices and it must also contain eleven θ's, because the r integration requires eleven r's to be non vanishing and all the terms with eleven r's also contain eleven θ's. There is only one invariant tensor, up to scaling, with these symmetries which is (ϵT):

$$\int d^{16}\theta f_{\alpha\beta\gamma}(\theta)(\epsilon T)^{\alpha\beta\gamma}_{\beta_1\cdots\beta_{11}}\theta^{\beta_1}\cdots\theta^{\beta_{11}}. \tag{2.66}$$

At higher loops two new issues arise, (1) appearance of the b ghost which is a composite field constructed from the worldsheet fields, including the non-minimal variables, (2) the weight one fields have zero modes. To deal with the second issue, $\mathcal{N}$ will also include zero modes of weight one fields. For the one-loop case weight one fields have one zero mode, this results in[4]

$$\mathcal{N}(y) = e^{-(\lambda(y)\bar{\lambda}(y)+r(y)\theta(y)+\frac{1}{2}N^0_{mn}\bar{N}^{mn}_0+\frac{1}{4}S^0_{mn}\int_A dz\lambda\gamma^{mn}d+J^0\bar{J}^0+S\int_A dz\lambda d)}. \tag{2.67}$$

This is invariant under Q_S:

$$Q_S\mathcal{N}(y) = (\lambda r(y) - \lambda r(y) + \bar{N}^{mn}\frac{1}{2}\lambda\gamma_{mn}d - \bar{N}^{mn}\frac{1}{2}\lambda\gamma_{mn}d + \bar{J}(\lambda d) - \bar{J}(\lambda d))\mathcal{N}(y) = 0. \tag{2.68}$$

The non-minimal b ghost satisfies

$$\{Q_S, b_{nm}(z)\} = T_{nm}(z) \equiv T_{min}(z) + T_{\bar{\lambda}\bar{w}}(z). \tag{2.69}$$

This equation can be solved in the non-minimal formalism and its solution is given by

$$b_{nm} = s^\alpha\partial\bar{\lambda}_\alpha + \frac{\bar{\lambda}_\alpha(2\Pi^m(\gamma_m d)^\alpha - N_{mn}(\gamma^{mn}\partial\theta)^\alpha - J\partial\theta^\alpha - \frac{1}{4}\partial^2\theta^\alpha)}{4\lambda\bar{\lambda}} \tag{2.70}$$

$$\frac{(\bar{\lambda}\gamma^{mnp}r)(d\gamma_{mnp}d + 24N_{mn}\Pi_p)}{192(\lambda\bar{\lambda})^2} - \frac{(r\gamma_{mnp}r)(\bar{\lambda}\gamma^m d)N^{np}}{16(\lambda\bar{\lambda})^3} + \frac{(r\gamma_{mnp}r)(\bar{\lambda}\gamma^{pqr}r)N^{mn}N_{qr}}{128(\lambda\bar{\lambda})^4}.$$

The one-loop amplitude prescription in the non-minimal pure spinor formalism is given by

$$\mathcal{A}^{(N)} = \langle V_1(z_1)\prod_{i=2}^{N}\int dz_i U_i(z_i)\int dw\mu(w)b_{nm}(w)\mathcal{N}(y)\rangle, \tag{2.71}$$

where $\mathcal{N}$ is given is (2.67). After integrating out the non zero modes by using the OPE's a typical one-loop amplitude in the non-minimal formalism becomes

$$\mathcal{A}^{(N)} = \int d^{16}dd^{16}\theta\int[d\lambda][d\bar{\lambda}][dN][d\bar{N}][ds][dr]f(\lambda,\bar{\lambda},\theta)\mathcal{N}^0, \tag{2.72}$$

where the Lorentz invariant measures are defined by

$$[d\bar{N}]\bar{\lambda}_{\alpha_1}\cdots\bar{\lambda}_{\alpha_8} = R^{m_1n_1\cdots m_{10}n_{10}}_{\alpha_1\cdots\alpha_8}d\bar{N}_{m_1n_1}\cdots d\bar{N}_{m_{10}n_{10}}d\bar{J} \tag{2.73}$$

and

$$[ds] = R^{\alpha_1\cdots\alpha_8}_{m_1n_1\cdots m_{10}n_{10}}\bar{\lambda}_{\alpha_1}\cdots\bar{\lambda}_{\alpha_8}\frac{\partial}{\partial S_{m_1n_1}}\cdots\frac{\partial}{\partial S_{m_{10}n_{10}}}\frac{\partial}{\partial S}. \tag{2.74}$$

[4]The zero mode of a holomorphic field $\phi(z)$ is given by: $\phi^0 \equiv \int_A dz\phi(z)$. A is the non-trivial A-cycle that satisfies $\int_A \omega(z) = 1$, where $w(z)$ is the holomorphic one-form on the torus.

Note b_{nm} has poles in $\lambda\bar{\lambda}$ which can cause the zero mode integrals over λ and $\bar{\lambda}$ to diverge. At one loop this will not cause any problems because the measure $[d\lambda][d\bar{\lambda}][dN][d\bar{N}][dr][ds]$ goes like $(\lambda)^{11}(\bar{\lambda})^{11}$ and the b field like $\bar{\lambda}/(\lambda\bar{\lambda})^4$ when $\lambda \to 0$. At three loops and higher the number of b fields is high enough to cause divergences. They have originally been regularised in [27] and more recently in [28], but this method has not been applied to actual computations. (See however [29] where this regularisation method is reviewed and applied to the one-loop four-point amplitude with four integrated vertex operators. This requires a modification of the amplitude prescription that will not be discussed in this thesis.)

2.3 Results from the pure spinor formalism

In this chapter two new string theory formalisms have been introduced. Although it is not been proved rigorously, there is a lot of evidence that the minimal pure spinor formalism, the non-minimal pure spinor formalism and the RNS formalism are equivalent to each other.

Let us start with the equivalence between RNS and the minimal pure spinor formalism. The spectra of these two were shown to coincide in[5] [24]. The most direct approach to show equivalence is to compare the amplitude computations. In [30] the equivalence was proved for N-point massless tree-level amplitudes with four or fewer Ramond states. For massless four-point one-loop amplitudes the amplitudes were shown to be identical in [31]. The four-point massless two-loop amplitude has been computed in the pure spinor formalism in [32]. This computation includes all possible choices (Neveu-Schwarz or Ramond) of the external states. The analogous computation in the RNS formalism is extremely complicated due to the sum over spin structures (cf. (1.213)) and is only successfully performed in the case of four NS states [17]. For this choice of external states the pure spinor result agrees with the RNS result. In conclusion one can say that the pure spinor formalism agrees with all known results of the RNS formalism. On top of this the pure spinor formalism produces more results, especially involving RR states, due to its manifest spacetime supersymmetry. All pure spinor computations referred to in this paragraph were performed in the manifestly Lorentz invariant version of the minimal formalism. This means including integrals over the constant tensors/spinors C and B (cf. (2.30)).

Equivalence of the minimal and non-minimal at tree-level is not difficult to show when one utilises the manifestly Lorentz invariant version of the minimal formalism.

[5]This reference shows coincidence of the spectrum of the minimal pure spinor formalism and the spectrum of the Green-Schwarz superstring, yet another superstring formalism. However as explained in [10] the GS string is equivalent to the RNS string.

The λ and C integrals in (2.30) can be evaluated by Lorentz invariance:

$$\int [dC][d\lambda]\lambda^\alpha\lambda^\beta\lambda^\gamma C^1_{\alpha_1}\cdots C^{11}_{\alpha_{11}}\delta(C^1\lambda)\cdots\delta(C^{11}\lambda) = (\epsilon T)^{\alpha\beta\gamma}_{\alpha_1\cdots\alpha_{11}}. \tag{2.75}$$

Using this result (2.30) becomes

$$\mathcal{A} = \int d^{16}\theta f_{\alpha\beta\gamma}(\theta)(\epsilon T)^{\alpha\beta\gamma}_{\alpha_1\cdots\alpha_{11}}\theta^{\alpha_1}\cdots\theta^{\alpha_{11}}, \tag{2.76}$$

which coincides with the non-minimal result (2.66). At higher loops there does not exist such a general proof, but in [33] the non-minimal one- and two-loop four-point functions are shown to coincide with their minimal counterparts. The most recent computation, the five-point one-loop amplitude, has only been computed in the non-minimal formalism [34]. In chapter 4 formal equivalence between the minimal and and non-minimal formalism will be proved by providing a first principles derivation from the same starting point for both minimal and non-minimal.

The power of the pure spinor formalism is not only illustrated by the fact that the complexity of all the amplitudes mentioned in the previous paragraph does not depend on the number of external fermions (unlike RNS). In addition there exists a number of non-renormalisation theorems that have been proved in the pure spinor formalism and not in RNS. Four theorems are listed below in chronological order. It is also indicated which formalism is used in the reference.

- The p-loop four graviton function vanishes above one loop [22] (minimal). In other words the R^4 term in the low energy effective action does not receive perturbative corrections above one loop. This is a consequence of a conjectured selfduality of type IIB string theory, S-duality. In the RNS formalism the conjecture was verified only at two loops after much effort [11].

- The massless N-point multiloop ($g \geq 2$) function vanishes whenever $N < 4$ [22] (minimal). This result is the main ingredient of the *proof of perturbative finiteness of string theory*. As explained in [22] the only other possible obstruction to proving perturbative finiteness is the existence of unphysical divergences in the interior of moduli space. Such divergences are not expected in the pure spinor formalism. Within the RNS formalism there are no results beyond two loops.

- In [35] (non-minimal) two more conjectures based on string dualities are presented and subsequently proved. The first theorem states that when $0 < n < 12$, $\partial^n R^4$ terms do not receive perturbative corrections above $n/2$ loops. The second theorem states that when $n \leq 8$, perturbative corrections to $\partial^n R^4$ terms in the IIA and IIB effective actions coincide.

Chapter 4

BRST quantisation of the pure spinor superstring

As mentioned throughout chapter 2 there are several unexplained aspects of the pure spinor formalism. These include

- the origin of the picture changing operators,
- the conformal weight constraint on the vertex operators,
- the b ghost equation (2.34),
- the relation between integrated and unintegrated vertex operators (2.17).

In a theory derived from first principles, for example the bosonic string, the above aspects all follow from one starting point, namely

$$Z = \int \mathcal{D}g\mathcal{D}X \frac{1}{VolG} e^{-S_P}. \tag{4.1}$$

In addition to providing an explanation for the above aspects a first principles derivation of the pure spinor formalism could also help in the search of a simplified version. Furthermore in chapter 2 it was advertised that one can replace all fields by their zero modes in a correlator that only contains weight zero fields. This will also be proved in this chapter.

In this chapter a first principles derivation is provided. There have been many works in the past involving modifications and/or extensions of the pure spinor formalism with the same aim, see for example [47, 48, 49, 50, 23, 51, 52, 53]. The approach of this chapter is different and is guided by topological string constructions. Instead of searching for a model with a local symmetry which after gauge fixing would lead to the pure spinor formalism with Q_S and the pure spinors emerging as a BRST operator and ghost fields, the pure spinors λ will be considered as "matter" fields as well and the worldsheet theory as a sigma model with a nilpotent symmetry Q_S and target space the ten-dimensional superspace times the pure spinor space. To construct a string theory this theory will be coupled to two-dimensional gravity in a way that preserves the fermionic symmetry Q_S and then BRST quantise the resulting theory in a conventional fashion.

4.1 Coupling to 2d gravity

To construct a string theory the pure spinor worldsheet action will be coupled to two-dimensional gravity in a way that preserves the Q_S symmetry. Subsequently this system will be quantised using BRST methods. Since this model has zero central charge, one should couple it to topological gravity[1]. This approach is thus similar to the construction of topological string theories, see [54] for a review. In that context one starts from a supersymmetric sigma model which upon topological twisting yields a topological sigma model. In this procedure one of the supersymmetry charges is identified with the BRST operator of the sigma model. The corresponding operator in the case at hand is the nilpotent operator Q_S. Note that the pure spinor sigma model has been obtained by twisting an $N = 2$ model in [23].

The first step in this procedure is thus to relax the conformal gauge in the action (2.1) (or (2.56) for the non-minimal version). The part that involves the x^m is standard[2],

$$S_X = \int d^2\sigma \left(\frac{1}{4}\sqrt{g}g^{ab}\partial_a x^m \partial_b x_m\right) \tag{4.2}$$

The rest of the action (2.1) (or (2.56) for the non-minimal version) is a sum of first order actions involving a field of dimension one and a field of dimension zero (with an overall sign that depends on whether the fields are bosonic or fermionic). The covariantisation of all these terms is the same, so it suffices to discuss one of them, say

$$S_{(p,\theta)} = \int d^2 z p_\alpha \bar{\partial}\theta^\alpha \ . \tag{4.3}$$

The fields of dimension one are vectors on the worldsheet, so p_α is more accurately labeled as $p_{a\alpha}$. However, only the z-component participates in (4.3), as one can conclude by looking at the conformal weight of the various objects in (4.3). Similarly, only the $\bar{z}$ component of the right-moving momentum $\tilde{p}_{a\alpha}$ participates in the action.

[1] By definition topological gravity does not change the central charge of the conformal field theory obtained after gauge fixing

[2] The worldsheet has a Euclidean signature and the conventions are the same as in chapter 1, i. e. $z = \sigma^1 + i\sigma^2$, the flat metric is $g_{z\bar{z}} = 1/2$ etc.

To account for this, one can introduce projection operators

$$P^{(\pm)b}_{\ a} = \frac{1}{2}(\delta_a{}^b \mp iJ_a{}^b), \tag{4.4}$$

where $J_a{}^b$ is the complex structure of the worldsheet, i.e. it satisfies

$$J_a{}^b J_b{}^c = -\delta^c_a, \qquad \nabla_c J_a{}^b = 0. \tag{4.5}$$

In terms of the worldsheet volume form and the worldsheet metric, it is given by $J_a{}^b = -\epsilon_{ac}g^{cb}$, with $\epsilon_{ab} = \sqrt{g}\hat{\epsilon}_{ab}$ and $\hat{\epsilon}_{01} = 1$, and holomorphic and anti-holomorphic functions on the worldsheet are defined by $J_a{}^b\partial_b f = i\partial_a f$ and $J_a{}^b\partial_b\tilde{f} = -i\partial_a\tilde{f}$, respectively. Using (4.5) one shows that

$$P^{(\pm)b}_{\ a}P^{(\pm)c}_{\ b} = P^{(\pm)c}_{\ a} \qquad P^{(\pm)b}_{\ a}P^{(\mp)c}_{\ b} = 0. \tag{4.6}$$

Notice also that

$$g^{ab}P^{(\pm)c}_{\ b} = g^{cb}P^{(\mp)a}_{\ b}. \tag{4.7}$$

One can obtain vectors with only z-component by multiplying by $P^{(+)b}_{\ a}$ and vectors with only $\bar{z}$-component by multiplying by $P^{(-)b}_{\ a}$:

$$\hat{p}_a = P^{(+)b}_{\ a}p_b, \qquad \hat{\tilde{p}}_a = P^{(-)b}_{\ a}\tilde{p}_b. \tag{4.8}$$

In other words, the only nonzero component of $P^{(+)b}_{\ a}$ is $P^{(+)z}_{\ z} = 1$ and the only nonzero component of $P^{(-)b}_{\ a}$ is $P^{(-)\bar{z}}_{\ \bar{z}} = 1$. More generally, these projection operators can be used to covariantise any tensor given in conformal gauge. The action (4.3) can then be covariantised as

$$S_{(p,\theta)} = \int d^2\sigma\sqrt{g}g^{ab}\hat{p}_{a\alpha}\partial_b\theta^\alpha. \tag{4.9}$$

In summary the action of the minimal model coupled to gravity is given by

$$S_\sigma = \int d^2\sigma\sqrt{g}g^{ab}\left(\frac{1}{4}\partial_a x^m\partial_b x_m + \hat{p}_{a\alpha}\partial_b\theta^\alpha - \hat{w}_{a\alpha}\partial_b\lambda^\alpha\right) \tag{4.10}$$

with an obvious addition for the case of the non-minimal model. The stress energy tensor for the model can be obtained by varying w.r.t. the worldsheet metric,

$$\begin{aligned} T_{ab} &= \frac{2}{\sqrt{g}}\frac{\delta S_\sigma}{\delta g^{ab}} = \frac{1}{2}(\partial_a x_m\partial_b x^m - \frac{1}{2}g_{ab}g^{cd}\partial_c x_m\partial_d x^m) \\ &+ (p_{(a|\alpha|}\partial_{b)}\theta^\alpha - \frac{1}{2}g_{ab}g^{cd}p_{c\alpha}\partial_d\theta^\alpha) + T^{(\lambda w)}_{ab} \end{aligned} \tag{4.11}$$

The contribution of the pure spinor part (and the non-minimal variables) is the same as the one for the (p,θ) part with $p \to w$ and $\theta \to \lambda$ and an overall minus sign (with similar replacements for the non-minimal fields). This stress energy tensor is (manifestly) traceless and covariantly conserved, reflecting the fact that the action is invariant under diffeomorphisms and Weyl transforms,

$$\begin{aligned}\delta g_{ab} &= \mathcal{L}_{\epsilon(\sigma)} g_{ab} + 2\phi(\sigma) g_{ab} \\ \delta\Phi &= -\epsilon^a \partial_a \Phi \\ \delta P_a &= -\epsilon^a \partial_a P + \partial_a \epsilon^b P_b\end{aligned} \tag{4.12}$$

where $\epsilon^a(\sigma), \phi(\sigma)$ are diffeomorphism and Weyl gauge parameters, $\mathcal{L}_\epsilon$ is the Lie derivative (cf. (1.65)), $\Phi = \{x^m, \theta^\alpha, \lambda^\alpha, \ldots\}$ collectively denotes all worldsheet scalars and $P_a = \{p_{a\alpha}, w_{a\alpha}, \ldots\}$ collectively denotes all worldsheet vectors.

The stress energy tensor (4.11) can be rewritten as

$$T_{ab} = P^{(+)c}_{\ \ a} P^{(+)d}_{\ \ b} T^B_{cd} + P^{(-)c}_{\ \ a} p_{c\alpha} \left(P^{(-)d}_{\ \ b} \partial_d \theta^\alpha \right) + \cdots \tag{4.13}$$

where the ellipsis indicate the contribution from the pure spinor and non-minimal variables, which will be suppressed from now on since they are similar to the (p, θ) contribution. The anti-holomorphic contribution of x^m is also suppressed. The first term in (4.13) is the covariantisation of the stress energy tensor appearing in Berkovits' work,

$$T^B_{ab} = \frac{1}{2} \partial_a x_m \partial_b x^m + p_{a\alpha} \partial_b \theta^\alpha + \cdots \tag{4.14}$$

while the second term is proportional to the θ^α field equation. This additional term can be removed by modifying the transformation rule of $p_{a\alpha}$ in (4.12).

4.1.1 Topological gravity and Q_S invariance

If one was to quantise the model just described one would find that it is anomalous, since the diffeomorphism ghosts would contribute $c = -26$ and the original sigma model had $c = 0$. This problem is avoided by extending the Q_S symmetry to act on the worldsheet metric, so that the $2d$ gravity is topological. With this aim, the following transformation rule is introduced,

$$\delta_S g_{ab} = P^{(-)c}_{\ \ a} P^{(-)d}_{\ \ b} \psi_{cd} \equiv \hat{\psi}_{ab}, \qquad \delta_S \hat{\psi}_{cd} = 0. \tag{4.15}$$

where ψ_{ab} is a new field that has only one holomorphic component, $\psi_{\bar{z}\bar{z}}(z)$. (To extend this discussion to the anti-holomorphic sector one would also need to turn on $\tilde{\psi}_{zz}(\bar{z})$, i.e. the full transformation is $\delta_S g_{ab} = P^{(-)c}_{\ \ a} P^{(-)d}_{\ \ b} \psi_{cd} + P^{(+)c}_{\ \ a} P^{(+)d}_{\ \ b} \tilde{\psi}_{cd}$).

Since the metric now transforms, the action is not invariant anymore and its Q_S variation yields,

$$\delta_S S_\sigma = -\frac{1}{2} \int d^2\sigma \sqrt{g}\, T^{ab} \delta_S g_{ab} = -\frac{1}{2} \int d^2\sigma \sqrt{g} g^{ac} g^{bd} T^B_{ab} \hat{\psi}_{cd}, \tag{4.16}$$

where again only the holomorphic sector is discussed, and the second equality makes use of the fact that due to the projector operators the second term in (4.13) does not contribute. To construct an invariant action one now has to add a new term to the action,

$$S_\sigma \to S = S_\sigma + \frac{1}{2}\int d^2\sigma\sqrt{g}g^{ac}g^{bd}G_{ab}\hat{\psi}_{cd} \tag{4.17}$$

The new action is invariant under the condition there exists a G_{ab} transforming as

$$\delta_S G_{ab} = T^B_{ab}. \tag{4.18}$$

Note that because $\hat{\psi}_{ab}$ has only one fermionic component, the variation of the explicit worldsheet metrics in the new term does not contribute. Including both sectors one finds that for the discussion to go through G_{ab} must be traceless. In conformal gauge and complex coordinates the (holomorphic part of) equation (4.18) becomes

$$T^B(z) = \{Q_S, G(z)\}. \tag{4.19}$$

The G currents generate a fermionic symmetry of the action in conformal gauge. In the language of [54] equation (4.19) defines the pure spinor action in conformal gauge to be a topological conformal theory.

Equation (4.18) for G_{ab} is the equation for a composite "b-field", cf. (2.34). Such a composite field has been constructed in conformal gauge in the non-minimal formalism. In the minimal case it was more difficult to solve equation (4.18). A detailed account of its solution will be given in section 4.4. Once the conformal gauge solution to (4.18) has been found, it can be covariantised to obtain a Q_S, diffeomorphism and Weyl invariant action.

4.2 Adding vertex operators

The vertex operators should be invariant under the symmetries of the theory, in this case: diffeomorphisms, Weyl transformations, Q_S transformations and the transformations generated by G_{ab}. In order to preserve the Q_S symmetry the vertex operators depend[3], in addition to the worldsheet coordinate σ_i^a, on its Q_S partner ζ_i^a,

$$\delta_S\sigma_i^a = \zeta_i^a, \qquad \delta_S\zeta_i^a = 0\,, \tag{4.20}$$

or in complex coordinates,

$$\delta_S z_i = \zeta_i, \quad \delta_S\bar{z}_i = \bar{\zeta}_i, \qquad \delta_S\zeta_i = 0, \quad \delta_S\bar{\zeta}_i = 0\,. \tag{4.21}$$

[3]It is not possible to choose $\delta_S\sigma_i^a = 0$, since the ζ's are needed to fix the residual gauge invariance of the symmetry generated by G_{ab}.

Since ζ_i is a fermionic variable the ith vertex operator V_i has the expansion (in complex basis)

$$V_i[\varphi](z_i, \zeta_i) = V_i^{(0)}[\varphi](z_i) + \zeta_i V_i^{(1)}[\varphi](z_i)\,, \tag{4.22}$$

where only the holomorphic part of the vertex operator is given. The symmetry generated by Q_S poses further constraints on the vertex operators:

$$\delta_S\left(V_i[\varphi](z_i, \zeta_i)\right) = 0\,. \tag{4.23}$$

The Q_S transformation can act either on worldsheet fields φ or on the positions z_i and we obtain

$$\delta_S V_i[\varphi](z_i, \zeta_i) = (\delta_S V_i^{(0)})(z_i) + \zeta_i\left(\partial V_i^{(0)}(z_i) - (\delta_S V_i^{(1)})(z_i)\right) \tag{4.24}$$

which implies

$$\delta_S V_i^{(0)} = 0, \qquad \delta_S V_i^{(1)} = \partial V_i^{(0)}\,, \tag{4.25}$$

where now Q_S acts only on the fields. The equality is exactly the relation between integrated and unintegrated vertex operators in the pure spinor formalism postulated in (2.17). Moreover from (4.25) one finds that the integrated vertex operator

$$U_i = \int dz V_i^{(1)} \tag{4.26}$$

is Q_S invariant.

The second transformation in (4.25) can be rewritten in a form that is useful to determine how G acts on the superfield components

$$\delta_S V_i^{(1)} = \delta_S\{G, V_i^{(0)}\}. \tag{4.27}$$

The partial derivative in (4.25) is generated by T and this can be replaced by a G transformation followed by a Q_S transformation. The G transformations of the components are given by

$$\{G, V_i^{(0)}\} = V_i^{(1)}, \quad [G, V_i^{(1)}] = 0. \tag{4.28}$$

Hence in order to construct a vertex operator invariant under the G symmetry one has to integrate over ζ:

$$\int d\zeta_i V_i[\varphi](z_i, \zeta_i). \tag{4.29}$$

Finally invariance under diffeomorphisms is achieved in the same way as in the bosonic string (cf. (1.78)), namely by integrating over the worldsheet coordinate z_i.

4.3 BRST quantisation

The action S in equation (4.17) constructed in the previous section is invariant under diffeomorphisms and local Weyl transformations. This theory can be quantised using the BRST methods developed in chapter 1. Recall that BRST quantisation amounts to adding a term $Q_V\Psi$ to the action, where Ψ is the gauge fixing fermion and Q_V is the BRST operator (cf. (1.51)). However in the case at hand there is a second nilpotent fermionic symmetry, generated by Q_S. In order to preserve both symmetries the gauge fixing term is of the following type

$$S \to S + \delta_V \delta_S \Psi. \tag{4.30}$$

In fact the order of the symmetries does not matter, since as shown in [54] Q_V and Q_S anticommute:

$$\{Q_V, Q_S\} = 0. \tag{4.31}$$

A proper gauge fixing condition for the group of diffeomorphism and Weyl transformation has been discussed at length in section 1.2.1 and the Q_S variation in (4.30) ensures that the G symmetry is also gauge fixed. The gauge fixing fermion has two terms in general, one that involves the metric (L_1) and one that involves vertex operators positions (L_2). The latter is only necessary on the sphere and on the torus, since only in those cases L_1 leaves residual gauge invariance, which can be fixed by imposing a condition on the vertex operator positions. The two gauge fixing terms are given by

$$L_1 = \delta_V \delta_S (\tilde{\beta}^{ab}[g_{ab} - \hat{g}_{ab}(\tau)]), \quad L_2 = \delta_V \delta_S \left(\sum_{\hat{j}=1}^{\kappa/2} \beta_a^{\hat{j}} (\sigma_{\hat{j}}^a - \hat{\sigma}_{\hat{j}}^a) \right), \tag{4.32}$$

where κ is the number of conformal killing vectors, $\hat{g}$ is the reference metric and $\hat{\sigma}$ are some chosen worldsheet positions. The β's are bosonic fields which can be concluded from the fact that the L's must be bosonic. Furthermore $\tilde{\beta}^{ab}$ is a tensor density such that L_1 is coordinate invariant. The object $\beta_a^{\hat{j}}$ does not depend on the worldsheet coordinates, it is similar to the $B_a^{\hat{i}}$'s from gauge fixing the residual gauge invariance in the bosonic string (cf.(1.80)), the only difference is the statistics.

The next step is performing the Q_S and Q_V transformations in the gauge fixing terms (4.32). To this end it is useful to have an overview of the transformations. The diffeomorphism and Weyl ghosts, c^a and C_ω have Q_S partners,

$$\delta_S c^a = \gamma^a, \qquad \delta_S C_\omega = \gamma_\omega\,, \tag{4.33}$$

which are bosonic BRST ghosts for the fermionic symmetry generated by G. Note that due to the nilpotency of both charges and the fact that they anticommute, the fields will appear in quartets. These quartets are given in figure 4.1.

$$
\begin{array}{ccccccccc}
\tau^k & \xrightarrow{\text{V}} & \xi^k & \tilde{\beta}^{ab} & \xrightarrow{\text{V}} & -p^{ab} & \sigma_i^a & \xrightarrow{\text{V}} & c^a(\sigma_i) \\
\text{S}\Big\downarrow & & \Big\downarrow\text{S} & \text{S}\Big\downarrow & & \Big\downarrow\text{S} & \text{S}\Big\downarrow & & \Big\downarrow\text{S} \\
\hat{\tau}^k & \xrightarrow{\text{-V}} & \hat{\xi}^k & \tilde{b}^{ab} & \xrightarrow{\text{-V}} & -B^{ab} & \zeta_i^a & \xrightarrow{\text{-V}} & -\gamma^a(\sigma_i)
\end{array}
$$

$$
\begin{array}{cccccc}
\beta_a^{\hat{j}} & \xrightarrow{\text{V}} & -p_a^{\hat{j}} & g_{ab} & \xrightarrow{\text{V}} & \mathcal{L}_c g_{ab} + 2C_\omega g_{ab} \\
\text{S}\Big\downarrow & & \Big\downarrow\text{S} & \text{S}\Big\downarrow & & \Big\downarrow\text{S} \\
b_a^{\hat{j}} & \xrightarrow{\text{-V}} & -B_a^{\hat{j}} & \psi_{ab} & \xrightarrow{\text{-V}} & \begin{array}{l}\mathcal{L}_c\hat{\psi}_{ab} - \mathcal{L}_\gamma g_{ab} \\ -2\gamma_\omega g_{ab} + 2C_\omega\hat{\psi}_{ab}\end{array}
\end{array}
$$

Figure 4.1: *Q_V and Q_S transformations on moduli, auxiliary/b ghost field, worldsheet coordinates, constant auxiliary/b ghost fields and the metric.*

Using the transformations in figure 4.1 one can process the gauge fixing terms in (4.32). Let us start with L_1:

$$
\begin{aligned}
L_1 &= \delta_V(\tilde{b}^{ab}[g_{ab} - \hat{g}_{ab}(\tau)] + \tilde{\beta}^{ab}[\hat{\psi}_{ab} - \hat{\tau}^k\partial_k\hat{g}_{ab}(\tau)] \qquad (4.34)\\
&= B^{ab}[g_{ab} - \hat{g}_{ab}(\tau)] - \tilde{b}^{ab}[2C_\omega g_{ab} + \mathcal{L}_c g_{ab} - \xi^k\partial_k\hat{g}_{ab}(\tau)] - p^{ab}[\hat{\psi}_{ab} - \hat{\tau}^k\partial_k\hat{g}_{ab}(\tau)] \\
&\quad + \tilde{\beta}^{ab}[\mathcal{L}_c\hat{\psi}_{ab} + 2C_\omega\hat{\psi}_{ab} - \mathcal{L}_\gamma g_{ab} - 2\gamma_\omega g_{ab} + \hat{\xi}^k\partial_k\hat{g}_{ab}(\tau) - \hat{\tau}^k\xi^l\partial_k\partial_l\hat{g}_{ab}(\tau)]\,,
\end{aligned}
$$

where $\partial_k\hat{g}_{ab}(\tau) = \partial\hat{g}_{ab}(\tau)/\partial\tau^k$ is a derivative of the reference metric w.r.t. the moduli and $\hat{\psi}_{ab}$ is defined in (4.15). This gauge fixing action contains the usual gauge fixing terms for the metric and the ghost actions for $\tilde{b}$, c and $\tilde{\beta}, \gamma$. The gauge fixing term for the residual gauge invariance can also be processed:

$$
\begin{aligned}
L_2 &= \delta_V\left(\sum_{\hat{j}=1}^{\kappa/2} b_a^{\hat{j}}(\sigma_{\hat{j}}^a - \hat{\sigma}_{\hat{j}}^a) + \beta_a^{\hat{j}}\zeta_{\hat{j}}^a\right) \qquad (4.35)\\
&= \sum_{\hat{j}=1}^{\kappa/2} B_a^{\hat{j}}(\sigma_{\hat{j}}^a - \hat{\sigma}_{\hat{j}}^a) - b_a^{\hat{j}}c^a(\sigma_{\hat{j}}) - p_a^{\hat{j}}\zeta_{\hat{j}}^a + \beta_a^{\hat{j}}\gamma^a(\sigma_{\hat{j}})\,.
\end{aligned}
$$

At this point all gauge symmetries have been treated, except the ones associated with zero modes of the original fields X, p, θ, w, λ. These will be discussed in the next section.

To summarise, a general scattering amplitude is given by

$$
Z = \int d\mu_\sigma d\mu \prod_{i=1}^{N} V_i[\varphi](\sigma_i, \zeta_i)\exp\left(-S - L_1 - L_2\right), \qquad (4.36)
$$

where S, L_1 and L_2 are given in (4.17),(4.34) and (4.35), $d\mu_\sigma$ is the measure factor associated with X, p, θ, w, λ (and non-minimal variables) that will be discussed in the next section and $d\mu$ is the measure that follows from the analysis of this section, i.e.

$$\begin{aligned} d\mu &= \prod_i^N d^2\sigma_i\sqrt{g(\sigma_i)}d^2\zeta_i \prod_{k=1}^{\mu} d\tau^k d\xi^k d\hat{\tau}^k d\hat{\xi}^k \prod_{\hat{\jmath}=1}^{\kappa/2} db_a^{\hat{\jmath}} dp_a^{\hat{\jmath}} d\beta_a^{\hat{\jmath}} dB_a^{\hat{\jmath}} \\ &\times \ \mathcal{D}\psi_{ab}\mathcal{D}g_{ab}\mathcal{D}c^a\mathcal{D}\gamma^a\mathcal{D}C_\omega\mathcal{D}\gamma_\omega\mathcal{D}p^{ab}\mathcal{D}\tilde{\beta}^{ab}\mathcal{D}B^{ab}\mathcal{D}\tilde{b}^{ab} \end{aligned} \tag{4.37}$$

The first line contains the integration over all constant "fields" while the fields in the second line are functionally integrated over. The integration over most of these variables can be done exactly.

As in previous sections only the holomorphic sector is discussed. Firstly, integrating over B^{ab} and g_{ab} sets the worldsheet metric equal to the reference metric $\hat{g}_{ab}$ in all expressions. Integrating over $B_a^{\hat{\jmath}}, p_a^{\hat{\jmath}}$, leads to delta functions $\delta(z_{\hat{\jmath}} - \hat{z}_{\hat{\jmath}})\delta(\zeta_{\hat{\jmath}})$ which can be used to integrate over $z_{\hat{\jmath}}, \zeta_{\hat{\jmath}}$. So $\kappa/2$ insertions will involve $V_{\hat{\jmath}}^{(0)}(\hat{z}_{\hat{\jmath}})$ while the remaining $(N - \kappa/2)$ vertex operators will involve $V_i^{(1)}(z_i)$ and will be integrated. Furthermore integrating out $b^{\hat{\jmath}}, \beta^{\hat{\jmath}}$ leads to the insertion $c(\hat{z}_{\hat{\jmath}})\delta(\gamma(\hat{z}_{\hat{\jmath}}))$.

Note that the $V_{\hat{\jmath}}^{(0)}$ and $V_i^{(1)}$ do not depend on the ghost fields[4], so the path integral factorises into a part that only depends on the ghosts and the rest. One might anticipate that the ghost contributions will cancel each other since c^a, C_ω and the γ^a, γ_ω are related by the Q_S symmetry. So to simplify the presentation the ghosts are set to zero. The complete computation including the ghosts is given in section 4.5. The scattering amplitudes thus take the form

$$\langle V_1 \cdots V_n\rangle = \int d\mu_\sigma e^{-S_\sigma} d\tilde{\mu} e^{-\tilde{S}} \prod_{\hat{\jmath}=1}^{\kappa/2} V_{\hat{\jmath}}^{(0)}(\hat{z}_{\hat{\jmath}}) \prod_{i=\kappa/2+1}^{N} \int dz_i V_i^{(1)}(z_i), \tag{4.38}$$

where

$$d\tilde{\mu}e^{-\tilde{S}} = \prod_{k=1}^{\mu} d\tau^k d\hat{\tau}^k \mathcal{D}\psi_{ab}\mathcal{D}p^{ab} \exp\left(\int d^2\sigma(\sqrt{\hat{g}}\frac{1}{2}G^{ab}\hat{\psi}_{ab} + p^{ab}[\hat{\psi}_{ab} - \hat{\tau}^k\partial_k\hat{g}_{ab}(\tau)])\right) \tag{4.39}$$

Integrating out p^{ab} gives a delta function that sets $\hat{\psi}_{ab} = \hat{\tau}^k\partial_k\hat{g}_{ab}(\tau)$. Finally integrating out $\hat{\tau}^k$ leads to $(6g-6)$ (of which $(3g-3)$ are holomorphic) insertions of G^{ab},

$$\langle V_1 \cdots V_n\rangle = \int d\mu_\sigma e^{-S_\sigma} \prod_k d\tau^k(G, \partial_k\hat{g}) \prod_{\hat{\jmath}=1}^{\kappa/2} V_{\hat{\jmath}}^{(0)}(\hat{z}_{\hat{\jmath}}) \prod_{i=\kappa/2+1}^{N} \int dz_i V_i^{(1)}(z_i) \tag{4.40}$$

[4]The ghost fields are consistently denoted as c/γ and lowercase b/β. In this case there are two b/β ghosts and four c/γ ghosts.

where $(G, \partial_k \hat{g}) = \int_\Sigma d^2\sigma \sqrt{\hat{g}} G^{ab} \partial_k \hat{g}_{ab}$.

4.3.1 Summary

Let us summarise the results so far. The starting point was a theory with a fermionic nilpotent symmetry Q_S and zero central charge. This theory was coupled to topological gravity in a way that preserves the Q_S symmetry. Quantising this system using BRST methods leads to the formula (4.40) for the scattering amplitudes. In this formula the position of $\kappa/2$ of the vertex operators $V_i^{(0)}$ is fixed while the remaining ones, $V_i^{(1)}$, are integrated. These vertex operators satisfy (in the holomorphic sector),

$$\delta_S V_i^{(0)} = 0, \qquad \delta_S V_i^{(1)} = \partial V_i^{(0)} . \tag{4.41}$$

Furthermore, one needs $(6g-6)$ insertions ($(3g-3)$ holomorphic ones) of the field G_{ab} defined by

$$\delta_S G_{ab} = T_{ab} \tag{4.42}$$

where T_{ab} is the stress energy tensor of the worldsheet theory. This composite field is the analogue of the b ghost in the scattering prescription of bosonic string theory. One may have anticipated these results based on the scattering amplitude prescription for the bosonic string and studies of topological strings. Indeed this is precisely the prescription used in the literature. The novelty here is its derivation from a first principles BRST quantisation. Notice that these results hold irrespectively of what the original sigma model is.

4.4 Pure spinor measure

Let us now return to the pure spinor sigma model. Two aspects deserve further attention. The first is finding an explicit form of the current G_{ab}. The second is determining whether the sigma model path integral measure $d\mu_\sigma$ contains gauge directions, i.e. whether evaluating the functional integral would lead to divergences. Let us start with the second one.

It turns out that there are gauge directions in $d\mu_\sigma$ and they are given by the zero modes of the sigma model (or matter) fields. The zero modes are gauge directions because by definition a zero mode is annihilated by the kinetic operator in the action and therefore zero modes do not appear in the action. For fermionic zero modes this does not present a problem; the vertex operators can provide the appropriate number of fermionic zero modes so that the final expressions are non-vanishing. Non-compact bosonic zero modes however are a problem, even in the presence of vertex operators. The action S_σ does not contain a convergence factor because of the zero mode gauge invariance. This can be remedied by gauge fixing the bosonic zero mode gauge invariances, as is discussed in this section. Due to the Q_S invariance, part of the invariance related to the fermionic zero modes is also fixed.

On a genus g surface, a worldsheet scalar Φ has one zero mode Φ_0 and a worldsheet vector P has g zero modes, $P_0(z) = \sum_{I=1}^{g} P^I \omega_I(z)$, where $\omega_I(z)$ are the g holomorphic Abelian differentials of first kind[5] satisfying $\int_{A_I} dz\omega_J = \delta_{IJ}$ and the contour integral is around the g non-trivial A-cycles of a genus g surface. Note that Φ_0 and P^I are constants. In the minimal pure spinor formalism there are ten zero modes x_0^m, sixteen zero modes θ_0^α and eleven zero modes λ_0^α from the worldsheet scalars and $16g$ zero modes $d_\alpha^I, I = 1, \dots g$, and $11g$ zero modes w_α^I from the worldsheet vectors. Of these x_0^m, λ_0^α and w_α^I are bosonic. The treatment of the zero modes of x^m is standard and will not be discussed here. Furthermore w_α, which transforms under the gauge transformation (2.4), will be traded for the gauge invariant variables,

$$N_{mn} = \frac{1}{2} w_\alpha (\gamma_{mn})^\alpha{}_\beta \lambda^\beta, \qquad J = w_\alpha \lambda^\alpha, \tag{4.43}$$

where N_{mn} is the (contribution of the pure spinors to the) Lorentz current and J is the ghost generator. As discussed in [27], the pure spinor condition implies enough relations between N_{mn} and J so that one can express the eleven independent components of w_α in terms of J and ten component of N_{mn}. In what follows the $11g$ zero modes of N_{mn}, J will be denoted by N_{mn}^I, J^I.

The zero mode gauge invariances cause divergences in the functional integral. Hence one can apply BRST quantisation to obtain a finite result. The BRST transformations corresponding to the zero mode gauge invariance are given by

$$\delta_V \lambda_0^\alpha = c^\alpha, \qquad \delta_V \theta_0^\alpha = \gamma^\alpha, \qquad \delta_V d_\alpha^I = \gamma_\alpha^I, \qquad \delta_V w_\alpha^I = c_\alpha^I, \tag{4.44}$$

where c^α, c_α^I are constant fermionic ghosts and $\gamma^\alpha, \gamma_\alpha^I$ are constant bosonic ghosts. The transformations for $\lambda_0^\alpha, w_\alpha^I$ require some explanation, since λ^α satisfy a quadratic constraint and w_α has a gauge invariance. These zero modes are most easily described in $U(5)$ variables since the system in terms of $\lambda^+, \lambda^{ab}, w_+, w_{ab}$ is unconstrained and has no gauge invariance (see section 3.3.2). The BRST transformation is then given by shifting these variables by their zero modes. Reversing the steps in section 3.3.2 one may express c^α in terms of the eleven zero modes of λ^+, λ^{ab} and c_α^I in terms of the $11g$ zero modes of w_+, w_{ab}. The arbitrariness due to the gauge invariance (2.4) is then eliminated by passing to the gauge invariant variables N_{mn}^I, J^I.

To maintain Q_S invariance one must further require

$$\delta_S \gamma^\alpha = c^\alpha, \qquad \delta_S c_\alpha^I = \gamma_\alpha^I. \tag{4.45}$$

[5]In the language of section 4.6, which contains a detailed account of what a zero mode really is, this Abelian differential is a realisation of $G_{0\kappa}$.

To gauge fix the bosonic invariances one needs constant fermionic and bosonic ghost fields, $b_\alpha, \tilde{b}_\alpha$ each containing eleven independent components, $\tilde{b}^{mnI}$, b^{mnI}, each containing $10g$ independent components and $\tilde{b}^I, b^I$, each containing g components and corresponding auxiliary fields. The Q_V and Q_S transformations of these fields are given in figure 4.2.

$$
\begin{array}{ccccccccc}
b_\alpha & \xrightarrow{V} & \pi_\alpha & \quad b^{mnI} & \xrightarrow{V} & \pi^{mnI} & \quad b^I & \xrightarrow{V} & \pi^I \\
\downarrow S & & \downarrow S & \quad \downarrow S & & \downarrow S & \quad \downarrow S & & \downarrow S \\
\tilde{b}_\alpha & \xrightarrow{-V} & \tilde{\pi}_\alpha & \quad \tilde{b}^{mnI} & \xrightarrow{-V} & \tilde{\pi}^{mnI} & \quad \tilde{b}^I & \xrightarrow{-V} & \tilde{\pi}^I
\end{array}
$$

Figure 4.2: *Q_V and Q_S transformations of the auxiliary and b ghost fields that play a role in gauge fixing zero mode invariances.*

The gauge fixing of the zero mode gauge invariances can be performed by introducing the following gauge fixing Lagrangian:

$$L_3 = \delta_V \delta_S \left(b_\alpha \theta_0^\alpha + \sum_{I=1}^{g} (b^{mnI} N_{mn}^I + b^I J^I) \right) \tag{4.46}$$

$$= \delta_V \left(-b_\alpha \lambda_0^\alpha + \tilde{b}_\alpha \theta_0^\alpha + \sum_{I=1}^{g} (\frac{1}{2} b^{mnI} (d^I \gamma_{mn} \lambda_0) + \tilde{b}^{mnI} N_{mn}^I + b^I (d^I \lambda_0) + \tilde{b}^I (w^I \lambda_0)) \right)$$

$$= -\pi_\alpha \lambda_0^\alpha - \tilde{\pi}_\alpha \theta_0^\alpha + \sum_{I=1}^{g} \left(\pi^{mnI} \frac{1}{2} d^I \gamma_{mn} \lambda_0 - \tilde{\pi}^{mnI} N_{mn}^I + \pi^I d_\alpha^I \lambda_0^\alpha - \tilde{\pi}^I J^I \right)$$

$$+ b_\alpha c^\alpha + \tilde{b}_\alpha \gamma^\alpha + \sum_{I=1}^{g} \left(\frac{1}{2} b^{mnI} (\gamma^I \gamma_{mn} \lambda_0 - d^I \gamma_{mn} c) - \frac{1}{2} \tilde{b}^{mnI} (c^I \gamma_{mn} \lambda_0 - w^I \gamma_{mn} c) \right.$$

$$\left. + b^I (\gamma^I \lambda_0 - d^I c) - \tilde{b}^I (c^I \lambda_0 - w^I c) \right).$$

Integrating over b^α and $\tilde{b}^\alpha$ leads to delta functions for c^α and γ^α, which can be used to integrate out c^α, γ^α. This sets c^α and γ^α to zero, in particular four of the eight terms in the sum in the last two lines of (4.46) disappear. Both the bosonic variables $b^{mnI}, b^I, \gamma_\alpha^I$ and the fermionic variables $\tilde{b}^{mnI}, \tilde{b}^I, c_\alpha^I$ only appear in the sum in the last lines of (4.46). Integration over these six variables leads to a factor of one, because the integration over the bosonic variables leads to a factor that is the inverse of the integral over the fermionic variables. More explicitly

$$\int [db^I][d\gamma^I] e^{\sum_I b^I \gamma^I \lambda_0} = \left(\int [d\tilde{b}^I][dc^I] e^{\sum_I \tilde{b}^I c^I \lambda_0} \right)^{-1}, \tag{4.47}$$

where the integration over b^{mnI} is suppressed to avoid cluttering of the equation. So the zero mode measure now becomes

$$[d\mu_\sigma]_{z.m.} = [d^{16}\theta_0][d^{11}\tilde{\pi}][d^{11}\lambda_0][d^{11}\pi] \prod_{I=1}^{g} [d^{11}\pi_I][d^{11}\tilde{\pi}_I][d^{11}N_I] \times \tag{4.48}$$

$$\exp\left(-\pi_\alpha \lambda_0^\alpha - \tilde{\pi}_\alpha \theta_0^\alpha + \sum_{I=1}^{g} \left(\pi^{mnI} \frac{1}{2} d^I \gamma_{mn} \lambda_0 - \tilde{\pi}^{mnI} N^I_{mn} + \pi^I d^I_\alpha \lambda_0^\alpha - \tilde{\pi}^I J^I \right) \right),$$

where $[d^{11}\lambda_0]$ and $\prod_I [d^{11}N_I]$ are the Lorentz invariant zero mode integration measures discussed in section 3.4, whose explicit form is not needed here. The auxiliary fields π seem to have to many components. For instance π_α has sixteen components whereas only eleven are needed to gauge fix the zero modes of λ. Similarly π_{mn} has 45 components while only ten are needed for the gauge fixing. This paradox can be resolved by realising that the exponent in (4.48) is invariant under a number of symmetries that render the "unwanted" components of π pure gauge. The symmetry for π_α is similar to the gauge invariance for w_α (cf. (2.4)):

$$\delta \pi_\alpha = f^m (\gamma_m \lambda)_\alpha. \tag{4.49}$$

This can be used to remove five components of π_a and since $\tilde{\pi}_\alpha = Q_S \pi_\alpha$ this propagates to $\tilde{\pi}_\alpha$. The symmetry for the higher loop auxiliary fields is given by

$$\delta \pi^I_{mn} = (\lambda \gamma_{[m})_\alpha f^{I\alpha}_{n]}, \qquad \delta \pi^I = -(\lambda \gamma^n f^I_n) \tag{4.50}$$

$$\delta \tilde{\pi}^I_{mn} = (\lambda \gamma_{[m})_\alpha \tilde{f}^{I\alpha}_{n]}, \qquad \delta \tilde{\pi}^I = -(\lambda \gamma^n \tilde{f}^I_n) \tag{4.51}$$

This symmetry can be used to eliminate 35 out of the 45 components of each π^I_{mn} and $\tilde{\pi}^I_{mn}$, which is as expected since the number of BRST auxiliary fields should be equal to the number of gauge fixing conditions.

The next step is actually integrating out $\pi, \tilde{\pi}, \pi^I, \tilde{\pi}^I$. This can be done in multiple ways, one leads to the minimal formalism and another to the non-minimal formalism.

4.4.1 Minimal formulation

The fields π_α and $\tilde{\pi}_\alpha$ have eleven independent components each. One way to parametrise them is to write

$$\pi_\alpha = p_i C^i_\alpha, \qquad \tilde{\pi}_\alpha = \tilde{p}_i C^i_\alpha, \quad i = 1, \ldots, 11 \tag{4.52}$$

where $p_i, \tilde{p}_i$ are the independent components and C^α_i is a constant matrix of rank eleven. Then $[d^{11}\pi][d^{11}\tilde{\pi}] = \prod_i dp_i d\tilde{p}_i$ and integrating over p^i yields $\prod_i \delta(C^i_\alpha \lambda^\alpha_0)$, while integrating over $\tilde{p}^i$ yields $\prod_i C^i_\alpha \theta^\alpha_0$. Putting it differently, one may have started with ghosts and auxiliary fields $b^i, \tilde{b}^i, p^i, \tilde{p}^i$ and gauge fixing condition $C^i_\alpha \lambda^\alpha_0 = 0$, for the invariance due to the eleven zero modes of λ^α and gauge fixing condition $C^i_\alpha \theta^\alpha_0 = 0$ for the invariance due to eleven of the sixteen zero modes of θ. Note that the insertions can be combined into eleven insertions of the picture lowering operator

$$Y_C = C_\alpha \theta^\alpha_0 \delta(C_\alpha \lambda^\alpha_0). \tag{4.53}$$

Similarly, one may parametrise the $10g$ independent components of π^{mnI} and of $\tilde{\pi}^{mnI}$ as

$$\pi^{mnI} = p^{jI} B^{mn}_{jI}, \qquad \tilde{\pi}^{mnI} = \tilde{p}^{jI} B^{mn}_{jI}, \qquad j = 1, \ldots, 10 \tag{4.54}$$

where $p^{jI}, \tilde{p}^{jI}$ are the $10g$ independent components and B^{mn}_{jI} are constants. Integrating over $p_{jI}, \tilde{p}_{jI}$ and $\pi_I, \tilde{\pi}_I$ leads to the insertions

$$\prod_{I=1}^{g} \left((d^I_\alpha \lambda^\alpha_0) \delta(J^I) \prod_{j=1}^{10} \frac{1}{2} B^{mn}_{Ij} (d^I \gamma_{mn} \lambda_0) \delta(B^{mn}_{Ij} N^I_{mn}) \right) = \prod_{R=1}^{g} Z_J(z_R) \prod_{P=1}^{10g} Z_{B_P}(w_P), \tag{4.55}$$

where the insertions have been reassembled in terms of the picture raising operators,

$$Z_B = \frac{1}{2} B^{mnI} d^I \gamma_{mn} \lambda_0 \delta(B^{mnI} N^I_{mn}), \qquad Z_J = (\lambda^\alpha_0 d^I_\alpha) \delta(J^I), \tag{4.56}$$

inserted at positions z_R, w_R. These insertions correspond to gauge fixing conditions $B^{mn}_{Ij} N^I_{mn} = 0, J^I = 0$, for the gauge invariance due to the $11g$ w_α zero modes and $B^{mn}_{Ij}(d^I \gamma_{mn} \lambda_0) = 0, d^I_\alpha \lambda^\alpha_0 = 0$ for the gauge invariance due to $11g$ of the $16g$ zero modes of d_α. Note that the constants C^i_α, B^{mn}_{Ij} enter through a gauge fixing term and there is a formal argument, presented below (1.51), that says physical predictions do not depend on the gauge fixing term and therefore not on B and C. However decoupling of Q_S exact states in the pure spinor formalism is a non trivial subject which is discussed in the next chapter. The precise statements about Lorentz invariance and dependence on B and C are specified there.

What is left is to discuss G_{ab}. By definition, G_{ab} should satisfy (now in complex coordinates and dropping the indices)

$$\delta_S G = T, \qquad T = \frac{1}{2} \Pi^m \Pi_m + d_\alpha \partial \theta^\alpha - w_\alpha \partial \lambda^\alpha. \tag{4.57}$$

Since δ_S is nilpotent, this equation defines a cohomology class $[G]$, i.e. solutions G up to δ_S exact terms. A solution of (4.57) is given by [55]

$$G_0 = \frac{C_\alpha G^\alpha}{C_\alpha \lambda^\alpha}, \qquad G^\alpha = \frac{1}{2} \Pi^m (\gamma_m d)^\alpha - \frac{1}{4} N_{mn} (\gamma^{mn} \partial \theta)^\alpha - \frac{1}{4} J \partial \theta^\alpha - \frac{1}{4} \partial^2 \theta^\alpha, \tag{4.58}$$

for a constant spinor C_α. This expression also appeared in [23] as a twisted worldsheet supersymmetry current. This solution is however not acceptable because it contains a factor of $(C_\alpha\lambda^\alpha)^{-1}$. Allowing such operators renders the Q_S cohomology trivial. Indeed, consider the field ξ

$$\xi = \frac{C_\alpha\theta^\alpha}{C_\alpha\lambda^\alpha}, \qquad \delta_S\xi = 1. \tag{4.59}$$

Then any closed operator V is also exact since

$$\delta_S V = 0 \quad \Rightarrow \quad V = \delta_S(\xi V). \tag{4.60}$$

A related issue is that the positions of the poles of G_0 are also the positions of the zeros of the path integral insertions thus making the expressions ill-defined.

One might hope to arrive at well-defined expression by finding a different representative of the cohomology class $[G]$ such that the poles in the new G cancel against zeros in other path integration insertions. Indeed, such a representative G_1 exists and it is given by $G_1 = b_B/Z_B$, where Z_B is the picture raising operator in (4.56) and b_B is the picture-raised b ghost originally constructed in [22] by solving the equation,

$$\delta_S b_B = Z_B T. \tag{4.61}$$

It was shown in [56] that G_1 is in the same cohomology class as G_0 and the poles of G_1 indeed cancel against zeros coming from the picture raising operators.

After the BRST quantisation the end result is that a multi-loop amplitude in the minimal pure spinor formalism should include $3g-3$ insertions of b_B, $10g-(3g-3)$ insertions of Z_B, g insertions of Z_J and eleven insertions of Y_C. This is precisely the prescription proposed in chapter 2.

4.4.2 Non-minimal formulation

Let us now return to (4.48) and recall that π_α and $\tilde{\pi}_\alpha$ are Q_S partners, $\delta_S\pi_\alpha = \tilde{\pi}_\alpha$, see figure 4.2, and each has eleven independent components. These are precisely the properties of the non-minimal variables $\bar{\lambda}_\alpha$ and r_α, see section 2, so one may identify

$$\pi_\alpha = \bar{\lambda}^0_\alpha, \qquad \tilde{\pi}_\alpha = r^0_\alpha \tag{4.62}$$

where $\bar{\lambda}^0_\alpha, r^0_\alpha$ are the zero modes of $\bar{\lambda}_\alpha$ and r_α. (Actually since the non-minimal variables are cohomologically trivial their nonzero modes do not contribute to any observable and one may only keep their zero modes). Recall also that the non-minimal sector has a gauge invariance similar to (2.4) (whose explicit form is not needed here) and the following combinations are gauge invariant [25]:

$$\bar{N}_{mn} = \frac{1}{2}(\bar{w}\gamma_{mn}\bar{\lambda} - s\gamma_{mn}r), \qquad \bar{J} = \bar{w}^\alpha\bar{\lambda}_\alpha - s^\alpha r_\alpha, \tag{4.63}$$

$$S_{mn} = \frac{1}{2}s\gamma_{mn}\bar{\lambda}, \qquad S = s^\alpha\bar{\lambda}_\alpha.$$

The canonical momenta $\bar{w}^\alpha$ and s^α have $11g$ zero modes each which, as in the discussion of the minimal variables, can be traded for $10g$ zero modes of $\bar{N}^I_{mn}$ and S^I_{mn} and g zero modes of $\bar{J}^I$ and S^I. Using the Q_S transformations in (2.58) one finds

$$\delta_S S^I_{mn} = \bar{N}^I_{mn}, \qquad \delta_S S^I = \bar{J}^I. \tag{4.64}$$

Thus the fields $\bar{N}^I_{mn}, S^I_{mn}, S^I, \bar{J}^I$ have the same degrees of the freedom and the same Q_S transformations as $\pi^{mnI}, \tilde{\pi}^{mnI}, \pi^I, \tilde{\pi}^I$. Therefore it is natural to identify them,

$$\pi^{mnI} = \bar{N}^{mnI}, \qquad \tilde{\pi}^{mnI} = S^I_{mn}, \qquad \pi^I = S^I, \qquad \tilde{\pi}^I = \bar{J}^I. \tag{4.65}$$

With these identifications the exponential factor in (4.48) is precisely the regularisation factor $\mathcal{N}$ of equation (2.67) (up to inconsequential numerical factors).

It remains to discuss G_{ab}. This field was constructed in [25] (with an elegant interpretation of the construction in terms of Čech cohomology given in [27])

$$G_B = \frac{\bar{\lambda}_\alpha G^\alpha}{(\bar{\lambda}\lambda)} + \frac{\bar{\lambda}_\alpha r_\beta H^{[\alpha\beta]}}{(\bar{\lambda}\lambda)^2} - \frac{\bar{\lambda}_\alpha r_\beta r_\gamma K^{[\alpha\beta\gamma]}}{(\bar{\lambda}\lambda)^3} - \frac{\bar{\lambda}_\alpha r_\beta r_\gamma r_\delta L^{[\alpha\beta\gamma\delta]}}{(\bar{\lambda}\lambda)^4}, \tag{4.66}$$

where G^α is given in (4.58) and $H^{\alpha\beta}, K^{\alpha\beta\gamma}, L^{\alpha\beta\gamma\delta}$ are specified in section 3.6. Note also that this field is cohomologically equivalent to G_0 [44]. Hence after a careful treatment of the zero mode invariances and finding the solution for G in the non-minimal formalism, the functional integral derived from first principles (4.40) reduces to the amplitude prescription advocated in section 2.2.

Notice that G_B field has poles as $\bar{\lambda}\lambda \to 0$ so one might wonder whether this prescription suffers from the same problems as the one using G_0. Indeed, there is a non-minimal version of the argument around (4.59)-(4.60). The corresponding non-minimal ξ field is [25]

$$\xi_{nm} = \frac{\bar{\lambda}_\alpha \theta^\alpha}{\bar{\lambda}_\beta \lambda^\beta + r_\beta \theta^\beta} \tag{4.67}$$

This diverges as $(\bar{\lambda}\lambda)^{-11}$ so one must ensure that no operators which diverge with this rate are allowed. A related issue is that the path integral with the insertions just discussed will diverge if the insertions diverge as fast as $(\bar{\lambda}\lambda)^{-11}$. As discussed in [25, 27] this can only happen for genus $g > 2$ (since the pure spinor measure converges as $(\bar{\lambda}\lambda)^{11}$ and G_B diverges as $(\bar{\lambda}\lambda)^{-3}$). One way to deal with this issue is look for a different representative $G_{(B,\epsilon)}$ of the Q_S cohomology class of $[G]$ which is less singular than G_B as $\bar{\lambda}\lambda \to 0$. A construction of such a $G_{(B,\epsilon)}$ is presented in [27]. Using this $G_{(B,\epsilon)}$ field one then arrives at a prescription that in principle works to all orders. See also [29] for more recent work.

This solves the problem in principle. The actual construction of $G_{(B,\epsilon)}$ however is very complicated. Given that the issues with singularities are related to the $\bar{\lambda}\lambda \to 0$ limit, a different approach would be to modify the gauge fixing condition for the pure spinor zero modes such that they are fixed to a nonzero value. It would be interesting to investigate if such gauge fixing can be implemented and whether it would lead to a simpler scattering amplitude prescription. Moreover chapter 5 will provide further motivation to look for a different gauge fixing condition.

4.5 Ghost contribution

There remains one loose end that needs to be tied up. In equation (4.38) the ghost fields were set to zero without sound motivation. This is provided in this section. Without setting the ghost fields to zero (4.38) is given by

$$\langle V_1 \cdots V_n \rangle = \int d\mu_\sigma e^{-S_\sigma} d\mu_{gh} e^{-S_{gh}} d\tilde{\mu} e^{-\tilde{S}} \prod_{\hat{j}=1}^{\kappa/2} V_{\hat{j}}^{(0)}(\hat{z}_{\hat{j}}) \prod_{i=\kappa/2+1}^{N} \int dz_i V_i^{(1)}(z_i), \tag{4.68}$$

where $d\tilde{\mu}e^{-\tilde{S}}$ is given in (4.39),

$$d\mu_{gh} = \mathcal{D}\tilde{\beta}^{ab}\mathcal{D}\tilde{b}^{ab}\mathcal{D}c^a\mathcal{D}\gamma^a\mathcal{D}C_\omega\mathcal{D}\gamma_\omega\mathcal{D}\xi^k\mathcal{D}\hat{\xi}^k \prod_{\hat{j}=1}^{\kappa} c^a(\hat{\sigma}_{\hat{j}})\delta(\gamma^a(\hat{\sigma}_{\hat{j}})) \tag{4.69}$$

and

$$\begin{aligned} S_{gh} = \int_\Sigma \Big(& 2\gamma_\omega\tilde{\beta}^{ab}\hat{g}_{ab}(\tau) - 2C_\omega(\tilde{b}^{ab}\hat{g}_{ab}(\tau) - \tilde{\beta}^{ab}\hat{\psi}_{ab}) \\ & + \tilde{b}^{ab}[\hat{\nabla}_a c_b + \hat{\nabla}_b c_a] + \tilde{\beta}^{ab}[\hat{\nabla}_a\gamma_b + \hat{\nabla}_b\gamma_a] + \tilde{b}^{ab}\xi^k\partial_k\hat{g}_{ab}(\tau) \\ & - \hat{\psi}_{ab}[\partial_c(\tilde{\beta}^{ab}c^c) - 2\tilde{\beta}^{c(b}\partial_c c^{a)}] - \tilde{\beta}^{ab}[\hat{\xi}^k\partial_k\hat{g}_{ab}(\tau) - \hat{\tau}^k\xi^l\partial_k\partial_l\hat{g}_{ab}(\tau)]\Big), \end{aligned} \tag{4.70}$$

where $\hat{\nabla}_a$ is the covariant derivative associated with $\hat{g}_{ab}$. The goal is to show that the "BRST factor" in (4.68), let us call it X_{BRST}, can be manipulated to give the result of section 4.3:

$$X_{BRST} = \int d\mu_{gh} e^{-S_{gh}} d\tilde{\mu} e^{-\tilde{S}} = \prod_{k=1}^{6g-6} d\tau^k(G, \partial_k\hat{g}(\tau)). \tag{4.71}$$

The first step is integrating out γ_ω and $\beta(\tau) \equiv \hat{g}_{ab}(\tau)\tilde{\beta}^{ab}$. This sets the trace of $\tilde{\beta}^{ab}$ equal to zero. The traceless part of $\tilde{\beta}^{ab}$ will be denoted by β^{ab}. Integrating out $\hat{\xi}^k$ introduces $(6g-6)$ insertions of the β^{ab} zero modes, while integrating over p^{ab}, ψ_{ab} and $\hat{\tau}^k$ leads to insertions of the zero mode of G,

$$(\tilde{G}, \partial_k\hat{g}) \equiv \int_\Sigma d^2\sigma \left(\partial_c(\beta^{ab}c^c) - 2\beta^{c(b}\partial_c c^{a)} + 2\beta^{ab}C_\omega + \sqrt{\hat{g}}G^{ab} + \beta^{ab}\xi^l\partial_l\right)\partial_k\hat{g}_{ab}(\tau). \tag{4.72}$$

After these integrations one is left with

$$X_{BRST} = \int d\mu_{\beta\gamma} d\tilde{\mu}_{gh} e^{-\tilde{S}_{gh}}, \tag{4.73}$$

where

$$\tilde{S}_{gh} = \int d^2\sigma \left(\beta^{ab}(\hat{\nabla}_a\gamma_b + \hat{\nabla}_b\gamma_a) + \tilde{b}^{ab}(2C_\omega\hat{g}_{ab} + \hat{\nabla}_a c_b + \hat{\nabla}_b c_a) + \tilde{b}^{ab}\xi^k\partial_k\hat{g}_{ab}(\tau)\right) \tag{4.74}$$

and

$$d\mu_{\beta\gamma} = [d\beta^{ab}][d\gamma^a]\prod_{k=1}^{6g-6}\delta((\beta,\partial_k\hat{g}))\prod_{\hat{\jmath}=1}^{\kappa}\delta(\gamma^a(\hat{\sigma}_{\hat{\jmath}})) \tag{4.75}$$

$$d\tilde{\mu}_{gh} = [d\tilde{b}^{ab}][dc^a][dC_\omega]\prod_{k=1}^{6g-6}d\tau^k(\tilde{G},\partial_k\hat{g}(\tau))\prod_{\hat{\jmath}=1}^{\kappa}c^a(\hat{\sigma}_{\hat{\jmath}}) \tag{4.76}$$

The $\beta\gamma$ system is now a standard CFT with a $U(1)$ charge conservation and the path integral measure contains all appropriate zero mode insertions. It follows that the β-dependent part of (4.72) drops out of (4.73) since it is charged w.r.t. the $\beta\gamma$ $U(1)$. Integrating out C_ω sets the trace of $\tilde{b}^{ab}$ to zero; the traceless part will denoted by b^{ab}, and integrating out ξ^k leads to $(6g-6)$ insertions of the b^{ab} zero modes. The BRST factor is now given by

$$X_{BRST} = \int d\mu_\tau d\mu_{\beta\gamma} d\mu_{bc} e^{-\int d^2\sigma\left(\beta^{ab}(\hat{\nabla}_a\gamma_b+\hat{\nabla}_b\gamma_a)+b^{ab}(\hat{\nabla}_a c_b+\hat{\nabla}_b c_a)\right)}, \tag{4.77}$$

with $d\mu_{\beta\gamma}$ as in (4.75) and

$$d\mu_{bc} = [db^{ab}][dc^a]\prod_{k=1}^{6g-6}(b,\partial_k\hat{g}(\tau))\prod_{\hat{\jmath}}^{\kappa}c^a(\hat{\sigma}_{\hat{\jmath}}), \quad d\mu_\tau = \prod_{k=1}^{6g-6}d\tau^k(G,\partial_k\hat{g}(\tau)). \tag{4.78}$$

It is now manifest that the integration over (b^{ab}, c^a) cancels against the integration over (β^{ab}, γ^a) and after integrating out b, c, β, γ one finds:

$$X_{BRST} = \prod_{k=1}^{6g-6}d\tau^k(G,\partial_k\hat{g}(\tau)). \tag{4.79}$$

4.6 Replacing worldsheet fields by zero modes

In chapter 2 the amplitude prescription for the pure spinor formalism was presented. In order to evaluate the correlators it was stated that one should first remove all fields of nonzero conformal weight by using the OPE's and thereafter replace the remaining fields, which all have weight zero, by their zero modes. The first principles derivation of this chapter has set the stage for justifying that step.

First it is useful to clarify what a zero mode really is. In general a zero mode is an eigenstate of some (differential) operator. In the case of $\beta\gamma$ systems it is not clear from the action in conformal gauge, i.e. for a special choice of coordinates, what this operator is, since such an action is only defined on one coordinate patch. Consider the $p\theta$ action, which is a fermionic $\beta\gamma$ system. Its action in conformal gauge is given by

$$S_{(p\theta)} = \int d^2 z p_\alpha \bar{\partial}\theta^\alpha, \tag{4.80}$$

where the weights of p and θ are one and zero respectively. In order to write this action in a coordinate free form, one uses the differential operators P_0 and P_0^T defined in (1.66). The action (4.80) can now be written in either of the two following forms

$$S_{(p\theta)} = (\hat{p}_\alpha, P_0\theta^\alpha) = (P_0^T \hat{p}_\alpha, \theta^\alpha), \tag{4.81}$$

where $(\hat{p}_\alpha)_a = P^{(+)b}_{\ \ a}(p_\alpha)_b$ and $P^{(+)b}_{\ \ a}$ is a projection operator defined in (4.4), not to be confused with the differential operator P_n, it depends on the complex structure of the worldsheet and in conformal gauge its only nonzero component is $P^{(+)z}_{\ \ z} = 1$. The operators P_n and P_n^T do not have eigenstates because they change the rank of the tensors they act on, therefore zero modes cannot be eigenstates of one of these operators. Operators that can be diagonalised are $P_n^T P_n$ and $P_n P_n^T$:

$$P_n P_n^T F_J^{a_1\cdots a_{n+1}}(\sigma) = v'^2_J F_J^{a_1\cdots a_{n+1}}(\sigma), \quad P_n^T P_n G_K^{a_1\cdots a_n}(\sigma) = v_K^2 G_K^{a_1\cdots a_n}(\sigma), \tag{4.82}$$

where F_J and G_K are symmetric traceless tensors of respectively rank $n+1$ and n. Any traceless symmetric worldsheet field can be expanded in a basis of eigenfunctions of $P_n^T P_n$ for some n. In addition it can expanded in eigenfunctions of $P_{n+1}P_{n+1}^T$. This basis can be chosen to be orthonormal with respect to (1.69):

$$(F_J, F_{J'}) = \delta_{JJ'}, \quad (G_K, G_{K'}) = \delta_{KK'}. \tag{4.83}$$

$\hat{p}_\alpha$ and θ^α can be expanded as

$$\hat{p}^a_\alpha(\sigma) = \sum_J (p_\alpha)_J \hat{F}^a_J(\sigma), \quad \theta^\alpha(\sigma) = \sum_K \theta^\alpha_K G_K(\sigma). \tag{4.84}$$

There is a one to one correspondence between the nonzero modes of $P_n^T P_n$ and $P_n P_n^T$. The number of zero modes can differ. This follows from

$$(P_n P_n^T) P_n G_J(\sigma) = P_n(P_n^T P_n) G_J(\sigma) = (v')^2_J P_n G_J(\sigma). \tag{4.85}$$

Thus $P_n G_J$ is an eigenfunction of $(P_n P_n^T)$. Along the same lines it follows that $P_n^T F_J$ is an eigenfunction of $(P_n^T P_n)$. This shows a one to one correspondence between the modes that satisfy $P_n G_J \neq 0$ and $P_n^T F_J \neq 0$, these are precisely the nonzero modes. These can be separated in positive $(J, K > 0)$ and negative modes $(J, K < 0)$. The zero modes are denoted as F_{0j}, G_{0k} where $j = 1, \cdots, \mu$ and $k = 1, \cdots, \kappa$. The values of μ and κ depend on both the genus of the worldsheet and the value of n. Canonical quantisation amounts to imposing the commutation relations

$$[F_J, G_K] = \delta_{J,-K}, \quad [F_J, F_{J'}] = [G_K, G_{K'}] = 0. \tag{4.86}$$

A vacuum can be defined by

$$F_J|0\rangle = G_K|0\rangle = 0, \quad J, K > 0. \tag{4.87}$$

Consider a number of $\beta\gamma$ systems of weight one, which is the relevant one for the pure spinor formalism. In the Hilbert space language a correlator, that only contains weight zero fields,

$$\mathcal{A} = \langle \gamma^1(z_1) \cdots \gamma^N(z_N) \rangle, \tag{4.88}$$

can be expanded as

$$\mathcal{A} = \langle 0|\gamma^1(z_1) \cdots \gamma^N(z_N)|0\rangle = \langle 0| \sum_{j=1}^{\kappa} (\gamma_0^1)_j G_{0j}(z_1) \cdots \sum_{j'=1}^{\kappa} (\gamma_0^N)_{j'} G_{0j'}(z_N)|0\rangle, \tag{4.89}$$

because the positive modes vanish against the vacuum on the right and the negative ones against the vacuum on the left. Also note there are only non vanishing (anti)commutators between a positive γ mode and a negative β mode or vice versa, so one can (anti)commute all γ modes through each other. This justifies replacing the fields in a correlator by their zero modes if all these fields are of weight zero.

Chapter 5

Decoupling of unphysical states

Any theory whose spectrum is defined as the cohomology of a certain nilpotent fermionic operator must have the property that all amplitudes with an exact state vanish. Otherwise two operators representing the same physical state give rise to different scattering amplitudes and there is no way to prefer either of the two answers. In chapter 2 it was shown that decoupling of unphysical states is guaranteed if all insertions are Q_S closed. For the non-minimal pure spinor formalism this was easy to show, cf. (2.61). The minimal formalism, however, contains constant spinors (C_α) and constant tensors (B_{mn}) in its amplitude prescription. These constant tensors enter the theory via the picture changing operators. It was argued in [22] that amplitudes are independent of C and B, because the Lorentz variation of the PCOs is Q_S exact.

In this chapter it will be shown by explicit computations that the amplitudes do depend on the choice of the constant tensors and Q_S exact states do not decouple. This happens already at tree level, but in this case one can show that there is a unique Lorentz invariant operator that can replace the PCOs in the tree-level amplitude prescription. With this replacement Q_S exact terms do decouple and one can further show that this prescription is equivalent to the tree-level prescription obtained by integrating over C [22], which correctly reproduces known tree-level amplitudes.

Amplitudes at one loop are discussed next. The main result will be that the PCOs, Y_C, are not Q_S closed. Furthermore a no-go theorem will be proved which states that Q_S closed Lorentz covariant PCOs lead to vanishing of all one-loop amplitudes. Hence if one wishes to replace the PCOs by Q_S closed ones, manifest Lorentz invariance cannot be maintained.

Note that $Q_S Y_C \neq 0$ by itself does not imply that Q_S exact states do not decouple. It only implies that the standard argument for decoupling of unphysical states that involves integrating Q_S by parts does not automatically lead to decoupling. Hence one needs another argument. This new argument does not use integration of Q_S by parts. Rather it makes use of an invariance of the path integral measure and the fact the zero mode integrals act as projectors on a certain Lorentz scalar. Then one can show that the integrand that results from Q_S exact insertions does not contain this scalar, hence amplitudes that contain unphysical states vanish after integration.

Even though there is a proof of decoupling of unphysical states in the formulation with integration over C and B, the fact that the PCOs are not Q_S closed is somewhat unsatisfactory. The technical origin of the problem is that the PCOs are Q_S closed only in a distributional sense and it turns out that the amplitudes are singular enough so that distributional identities do not hold. To understand why the amplitudes are singular, let us recall that the PCOs originate from gauge fixing zero mode invariances as discussed in chapter 4. The PCOs contain eleven delta functions of the form $\delta(C^I_\alpha \lambda^\alpha)$, where C^I_α are the constant spinors mentioned above. It turns out that for any choice of C^I that give an irreducible set of eleven constraints, the solution of $C^I_\alpha \lambda^\alpha = 0$ is given by $\lambda^\alpha = 0$, which is the tip of the cone that represents pure spinor space. As discussed in [21], the $\lambda^\alpha = 0$ locus should be removed from the pure spinor space. Thus this prescription corresponds to a singular gauge fixing condition and the problems with Q_S closedness of the PCOs reflect that fact. Furthermore the PCOs are not globally defined on pure spinor space. Ultimately one would like to use globally defined, Q_S closed PCOs that gauge fix the zero modes of λ to a nonzero value. Such an operator has not been found. Note however that this operator cannot be a Lorentz scalar, due to the no-go theorem.

There is one final point that deserves to be mentioned in this introduction. As stated in chapter 2 the most complicated loop amplitude computations have only been performed in the non-minimal pure spinor formalism. This suggests that the minimal loop computations are technically more involved. The analysis of this chapter, in particular the previously unnoted invariance of the path integral measure, might be used to simplify minimal loop computations.

5.1 Tree level

In the first part of this section, a number of tree-level amplitudes is computed in the formulation without an integral over the constant spinors C. The conclusion will be that these amplitudes are not Lorentz invariant and unphysical states do not decouple. In the second part a manifestly Lorentz invariant prescription without constant spinors is presented. As will be shown this new prescription leads to decoupling of unphysical states and is equivalent to the prescription with an integral over the constant spinors C.

5.1.1 No C integration

This section presents two problems regarding the minimal amplitude prescription (2.22) when it is evaluated using the definition of the zero mode measure (2.25) and the usual definition of a delta function:

$$\int dx\delta(x)f(x) = f(0), \quad x\delta'(x) = -\delta(x). \tag{5.1}$$

The problems are

- $\mathcal{A}$ is not Lorentz invariant or equivalently $\mathcal{A}$ depends on the choice of C's
- Q_S exact states do not decouple.

Lorentz invariance

In section 2.1.2 it was argued the PCOs are Lorentz invariant inside correlators if they are Q_S closed. The Q_S variation is given in (2.29) and this seems to vanish but if one chooses $C_\alpha = \delta_\alpha^+$, the result is $Q_S Y_C = \lambda^+\delta(\lambda^+)$. This is not zero because the measure contains $\frac{1}{(\lambda^+)^3}$. All one can use is $(\lambda^+)^4\delta(\lambda^+) = 0$. This problem is made even more explicit in the computation below. It will be shown that choosing particular C's does not result into a Lorentz invariant answer.

Let us choose[1]

$$C^1_\alpha = \delta^+_\alpha, \quad (C^2)^{a_1a_2} = \delta_1^{[a_1}\delta_2^{a_2]}, \ldots, (C^{11})^{a_1a_2} = \delta_4^{[a_1}\delta_5^{a_2]}, \quad \text{all other } C^I_\alpha = 0. \tag{5.2}$$

Note C^I_α has rank eleven for this choice, as it should. As is discussed in section 5.5 the lack of Lorentz invariance, which is shown below, would also be found, if any other choice was made, see footnote 9. The three-point tree-level function is given by

$$\mathcal{A} = \langle \lambda^\alpha A_{1\alpha}(z_1)\lambda^\beta A_{2\beta}(z_2)\lambda^\gamma A_{3\gamma}(z_3)Y_{C_1}(\infty)\cdots Y_{C_{11}}(\infty)\rangle. \tag{5.3}$$

The PCOs operators are inserted at infinity, since this simplifies the computation. All OPE's of the PCOs with the vertex operators vanish due to this choice. Therefore one can replace all fields in (5.3) by their zero modes:

$$\begin{aligned}\mathcal{A} &= \int [d\lambda]d^{16}\theta\lambda^\alpha\lambda^\beta\lambda^\gamma f_{\alpha\beta\gamma}(\theta)C^1_{\alpha_1}\theta^{\alpha_1}\cdots C^{11}_{\alpha_{11}}\theta^{\alpha_{11}}\delta(C^1_{\alpha_1}\lambda^{\alpha_1})\cdots\delta(C^{11}_{\alpha_{11}}\lambda^{\alpha_{11}}) \quad (5.4)\\ &= \int [d\lambda]d^{16}\theta\lambda^\alpha\lambda^\beta\lambda^\gamma f_{\alpha\beta\gamma}(\theta)\theta^+\theta_{12}\cdots\theta_{45}\delta(\lambda^+)\delta(\lambda_{12})\cdots\delta(\lambda_{45})\end{aligned}$$

[1]See section 3.2.1 for notational conventions.

$$= \int \frac{d\lambda^+ \wedge d\lambda_{12} \wedge \cdots \wedge d\lambda_{45}}{\lambda^{+3}} d^{16}\theta \lambda^\alpha \lambda^\beta \lambda^\gamma f_{\alpha\beta\gamma}(\theta)\theta^+\theta_{12}\cdots\theta_{45}\delta(\lambda^+)\delta(\lambda_{12})\cdots\delta(\lambda_{45}).$$

The only term that contributes is the one with $\alpha\beta\gamma = +++$, in all other cases there is an integral of the form $\int d\lambda_{ab}\lambda_{ab}\delta(\lambda_{ab})$ (no sum). There is a subtlety with these integrals, for instance

$$\int [d\lambda](\lambda^+)^2\lambda_{cd}\delta(\lambda^+)\delta(\lambda_{12})\cdots\delta(\lambda_{45}) = \int d\lambda^+ d^{10}\lambda_{ab}\frac{\lambda_{cd}}{\lambda^+}\delta(\lambda^+)\delta(\lambda_{12})\cdots\delta(\lambda_{45}) =$$

$$\int d\lambda^+ \frac{1}{\lambda^+}\delta(\lambda^+) \int d\lambda_{cd}\lambda_{cd}\delta(\lambda_{cd}) = \infty 0. \tag{5.5}$$

Note however that (5.5) has N charge one (cf. (3.50)). Since the outcome of the integral (maybe after some regularisation) must be a number, which does not transform under N, the integral has to vanish. In other words only integrals with zero N charge, like $\int [d\lambda](\lambda^+)^3\delta(\lambda^+)\delta(\lambda_{12})\cdots\delta(\lambda_{45})$ can be non-vanishing. After the integration over the λ zero modes one is left with

$$\mathcal{A} = \int d^{16}\theta f_{+++}\theta^+\theta_{12}\cdots\theta_{45}, \tag{5.6}$$

where $f_{+++} = A^1_+ A^2_+ A^3_+$ and this can be evaluated with the help of the explicit expressions for the gamma matrices from section 3.2.4. If one chooses the external states to be two gauginos and one gauge boson the amplitude becomes:

$$\mathcal{A} = \int d^{16}\theta(\xi^a_1\theta_{ka}\theta^k + \xi^1_{ka}\theta^a\theta^k)(\xi^b_2\theta_{lb}\theta^l + \xi^2_{lb}\theta^b\theta^l)\theta^c a^3_c\theta^+\theta_{12}\cdots\theta_{45} = \epsilon^{abcde}\xi^1_{ab}\xi^2_{cd}a^3_e. \tag{5.7}$$

This answer is not Lorentz invariant and different from the expected answer,

$$\xi^1\gamma^m\xi^2 a^3_m = 2(\xi^+_1\xi^a_2 a^3_a + \xi^a_1\xi^+_2 a^3_a - \frac{1}{4}\epsilon^{abcde}\xi^1_{ab}\xi^2_{cd}a^3_e + \xi^1_{ab}\xi^a_2 a^b_3 + \xi^a_1\xi^2_{ab}a^b_3), \tag{5.8}$$

where m is an $SO(10)$ index and all Latin letters that come before m in the alphabet are $SU(5)$ indices. In conclusion this shows that tree-level amplitudes do not yield Lorentz invariant answers when one does not integrate over C.

Dependence on C^I

On top of the lack of Lorentz invariance amplitudes depend on the choice of constant spinors C. In other words they are not invariant under $C^I_\alpha \to C^I_\alpha + \delta C^I_\alpha$. This variation changes the Ith PCO by a Q_S exact quantity. However when one computes a tree-level amplitude with the Ith PCO replaced by this Q_S exact quantity, it does not vanish. Hence incidentally this computation demonstrates that not all Q_S exact states decouple. In the computation below the same C's as in (5.2) are used and

$\delta C^{11}_\alpha = \delta^1_\alpha$, where the 1 is an $SU(5)$ vector index. The delta only has one non-vanishing component. This changes $Y_{C_{11}}$ by

$$\begin{aligned}\delta Y_{C_{11}} &= \delta C_{11\alpha}\theta^\alpha \delta(C_{11}\lambda) + C_{11\alpha}\theta^\alpha \delta C_{11\beta}\lambda^\beta \delta'(C_{11}\lambda) \\ &= Q_S(\delta C_{11\alpha}\theta^\alpha C_{11\beta}\theta^\beta \delta'(C_{11\beta}\lambda^\beta)) = Q_S(\theta^1\theta_{45}\delta'(\lambda_{45})).\end{aligned} \tag{5.9}$$

Under this change in C^I_α the tree-level three-point function changes by

$$\begin{aligned}\delta\mathcal{A} &= \langle V_1(z_1)V_2(z_2)V_3(z_3)Y_{C_1}(\infty)\cdots Y_{C_{10}}(\infty)\delta Y_{C_{11}}(\infty)\rangle = \\ &\langle V_1(z_1)V_2(z_2)V_3(z_3)Q_S(Y_{C_1}(\infty)\cdots Y_{C_{10}}(\infty))\theta^1(\infty)\theta_{45}(\infty)\delta'(\lambda_{45}(\infty))\rangle \\ &= \int d^{16}\theta \frac{d^{11}\lambda}{(\lambda^+)^3}\lambda^\alpha\lambda^\beta\lambda^\gamma A^1_\alpha A^2_\beta A^3_\gamma Q_S(Y_{C_1}\cdots Y_{C_{10}})\theta^1\theta_{45}\delta'(\lambda_{45}).\end{aligned} \tag{5.10}$$

There is a total of four λ^α's in the numerator (one hidden in Q_S) one of them has to be λ_{45} and the other three have to be λ^+ to give a non-vanishing answer. The term that contributes comes from Q_S hitting $\theta^+\delta(\lambda^+)$, this λ^+ then cancels against a λ^+ in the denominator and the variation becomes

$$\begin{aligned}\delta\mathcal{A} &= \int d^{16}\theta d^{11}\lambda A^{(1}_+ A^2_+ (A^{3)})^{45}\theta^1\delta(\lambda^+)\theta_{12}\delta(\lambda_{12})\cdots\theta_{45}\delta(\lambda_{45}) = \\ &\int d^{16}\theta A^{(1}_+ A^2_+ (A^{3)})^{45}\theta^1\theta_{12}\cdots\theta_{45}.\end{aligned} \tag{5.11}$$

By choosing suitable polarisations it is not difficult to see this does not always vanish.

5.1.2 Including C integration

Obtaining amplitudes which are not Lorentz invariant is a serious problem and one might ask why the tree-level amplitude computations [22, 57] in the minimal pure spinor formalism gave Lorentz invariant answers and why Q_S exact states decoupled. Both these points are explained in the first part of this section. In the second part the tree-level amplitude prescription is reformulated in a way that does not contain any constant spinors.

Lorentz invariance is restored by integrating over all possible choices of C^I_α, and this also results in decoupling of Q_S exact states as will become apparent in this section. The manifestly Lorentz invariant tree-level amplitude in the minimal formalism is given by

$$\mathcal{A} = \int [dC]\langle V_1(z_1)V_2(z_2)V_3(z_3)\int dz_4 U_4(z_4)\cdots\int dz_N U_N(z_N)Y_{C_1}(\infty)\cdots Y_{C_{11}}(\infty)\rangle. \tag{5.12}$$

After performing the OPE's and replacing the fields by their zero modes this becomes

$$\mathcal{A} = \int [dC]\int [d\lambda]d^{16}\theta\lambda^\alpha\lambda^\beta\lambda^\gamma f_{\alpha\beta\gamma}(\theta)(C^1\theta)\delta(C^1\lambda)\cdots(C^{11}\theta)\delta(C^{11}\lambda). \tag{5.13}$$

Note the eleven PCOs, Y_C, have been replaced by a manifestly Lorentz invariant PCO which will be called Y:

$$Y \equiv \int [dC](C^1\theta)\delta(C^1\lambda)\cdots(C^{11}\theta)\delta(C^{11}\lambda). \tag{5.14}$$

Now one uses

$$\int [dC][d\lambda]\lambda^\alpha\lambda^\beta\lambda^\gamma C^1_{\beta_1}\cdots C^{11}_{\beta_{11}}\delta(C^1\lambda)\cdots\delta(C^{11}\lambda) = (\epsilon T)^{\alpha\beta\gamma}_{\beta_1\cdots\beta_{11}}. \tag{5.15}$$

This is justified by Lorentz invariance, because the LHS is Lorentz invariant and the only invariant tensor with the appropriate symmetries is[2] (ϵT), as can be verified with [40]. Thus

$$\mathcal{A} = (\epsilon T)^{\alpha\beta\gamma}_{\alpha_1\cdots\alpha_{11}} \int d^{16}\theta f_{\alpha\beta\gamma}(\theta)\theta^{\alpha_1}\cdots\theta^{\alpha_{11}}. \tag{5.17}$$

The amplitude $\mathcal{A}$ is manifestly Lorentz invariant.

This prescription also ensures the decoupling of unphysical states. Amplitudes with unphysical states will be denoted by $\mathcal{B}$ throughout this chapter, while $\mathcal{A}$ is used for any amplitude, so at tree level with $V_1 = Q_S\Omega$,

$$\mathcal{B} = \int [dC]\langle Q_S\Omega(z_1)V_2(z_2)V_3(z_3)\prod_{i=4}^{N}\int dz_i U(z_i) \tag{5.18}$$

$$C^1_{\alpha_1}\theta^{\alpha_1}\cdots C^{11}_{\alpha_{11}}\theta^{\alpha_{11}}\delta(C^1\lambda)\cdots\delta(C^{11}\lambda)\rangle.$$

This can be written in the following form:

$$\mathcal{B} = \int [dC]\langle\lambda^\alpha(z_2)\lambda^\beta(z_3)g_{\alpha\beta}(d,\theta,N)Q_S(C^1_{\alpha_1}\theta^{\alpha_1}\cdots C^{11}_{\alpha_{11}}\theta^{\alpha_{11}})\delta(C^1\lambda)\cdots\delta(C^{11}\lambda)\rangle \sim$$

$$\int [dC]\langle\lambda^\alpha(z_2)\lambda^\beta(z_3)g_{\alpha\beta}(d,\theta,N)C^1_{\alpha_1}\lambda^{\alpha_1}\cdots C^{11}_{\alpha_{11}}\theta^{\alpha_{11}}\delta(C^1\lambda)\cdots\delta(C^{11}\lambda)\rangle. \tag{5.19}$$

where in going from the first to the second line an overall numerical factor of eleven was omitted. Such overall inconsequential factors will be neglected throughout this

[2]Incidentally, the following related integral can also be computed using Lorentz invariance:

$$\int [dC]d\lambda^{\alpha_1}\wedge\cdots\wedge d\lambda^{\alpha_{11}}C^1_{\beta_1}\cdots C^{11}_{\beta_{11}}\delta(C^1\lambda)\cdots\delta(C^{11}\lambda) = \tag{5.16}$$

$$c_1\delta^{[\alpha_1}_{\beta_1}\cdots\delta^{\alpha_{11}]}_{\beta_{11}} + c_2\gamma^{[\alpha_1\alpha_2}_{mnp}\gamma^{mnp}_{[\beta_1\beta_2}\delta^{\alpha_3}_{\beta_3}\cdots\delta^{\alpha_{11}]}_{\beta_{11}]},$$

where c_1 and c_2 are nonzero numerical constants. This structure follows from the fact $\text{Asym}^{11}\mathbf{16}\otimes\text{Asym}^{11}\mathbf{16}'$ contains two scalars (see section 3.2.2 for explanation about the notation and the argument). The constants can be computed using judicious choices of the indices. For example, the integral vanishes for the choice $\alpha_1 = \beta_1, \cdots, \alpha_{11} = \beta_{11} = +, 12, \ldots, 35, 5$, implying that one needs a nonzero constant c_2. Equation (5.16) corrects formula (3.25) of [22].

chapter. After using the OPE's to integrate out the nonzero modes one gets:

$$\begin{aligned}\mathcal{B} &= \int [dC] d^{16}\theta [d\lambda] \lambda^\alpha \lambda^\beta f_{\alpha\beta}(\theta) C^1_{\alpha_1}\lambda^{\alpha_1} C^2_{\alpha_2}\theta^{\alpha_2}\cdots C^{11}_{\alpha_{11}}\theta^{\alpha_{11}}\delta(C^1\lambda)\cdots\delta(C^{11}\lambda) = \\ &\int d^{16}\theta f_{\alpha\beta}(\theta)(\epsilon T)^{\alpha\beta\alpha_1}_{\alpha_1\cdots\alpha_{11}}\theta^{\alpha_2}\cdots\theta^{\alpha_{11}} = 0, \end{aligned} \tag{5.20}$$

where $f_{\alpha\beta}(\theta)$ is some function of θ zero modes and (5.15) was used in the second equality. The integral vanishes because[3] $\mathbf{126} \otimes \mathrm{Asym}^{10}\mathbf{16}$ does not contain a scalar (see section 3.2.2 for explanation about the notation and the argument), in other words

$$(\epsilon T)^{\beta_1\beta\gamma}_{\beta_1\cdots\beta_{11}} = 0. \tag{5.21}$$

In this case one can also write out (ϵT) explicitly and check that its trace contains a contraction of an antisymmetric tensor (ϵ) and a symmetric one $(\gamma^{\alpha\beta}_m)$.

Lorentz invariant tree-level prescription without constant spinors

There exists a replacement for the eleven PCOs that does not contain any constant spinors and is manifestly Lorentz covariant. The prescription that uses this replacement is equivalent to the one given in [22], when the integral over C in included. The prescription is given by

$$\mathcal{A} = \langle V_1(z_1)V_2(z_2)V_3(z_3)\int dz_4 U_4(z_4)\cdots\int dz_N U_N(z_N)\Lambda_{\alpha\beta\gamma}(\infty) \tag{5.22}$$

$$(\epsilon T)^{\alpha\beta\gamma}_{\beta_1\cdots\beta_{11}}\theta^{\beta_1}(\infty)\cdots\theta^{\beta_{11}}(\infty)\rangle.$$

The replacement of the eleven PCOs Y_C is called $\Lambda_{\alpha\beta\gamma}(\infty)$. After integrating out the nonzero modes and replacing the fields by their zero modes $\mathcal{A}$ reduces to

$$\mathcal{A} = \int d^{16}\theta [d\lambda]\lambda^\alpha\lambda^\beta\lambda^\gamma f_{\alpha\beta\gamma}(\theta)(\epsilon T)^{\delta_1\delta_2\delta_3}_{\beta_1\cdots\beta_{11}}\theta^{\beta_1}\cdots\theta^{\beta_{11}}\Lambda_{\delta_1\delta_2\delta_3}. \tag{5.23}$$

The tensor $\Lambda_{\alpha\beta\gamma}$ is defined by

$$\int [d\lambda]\lambda^\alpha\lambda^\beta\lambda^\gamma\Lambda_{\alpha'\beta'\gamma'} = \delta^{(\alpha}_{\alpha'}\delta^{\beta}_{\beta'}\delta^{\gamma)}_{\gamma'} - \frac{1}{40}\gamma^{(\alpha\beta}_m\gamma^m_{(\alpha'\beta'}\delta^{\gamma)}_{\gamma')} \equiv \delta^{((\alpha}_{\alpha'}\delta^{\beta}_{\beta'}\delta^{\gamma))}_{\gamma'}, \tag{5.24}$$

and is a function of the λ's only. More accurately, all components contain eleven delta functions or derivatives thereof. The precise form of (5.24) follows from the fact that the integral must be an invariant tensor combined with the pure spinor constraint. Detailed arguments are provided in section 3.5. Explicit expressions of the components can be found by examining certain components of (5.24). In order

[3] Note **126** denotes a gamma matrix traceless symmetric rank two tensor (recall that $\lambda^\alpha\lambda^\beta \sim \lambda\gamma^{mnpqr}\lambda\gamma^{\alpha\beta}_{mnpqr}$).

to see what conditions (5.24) imposes on Λ_{+++} note that choosing $\alpha\beta\gamma = +++$ gives

$$\int [d\lambda] {\lambda^+}^3 \Lambda_{+++} = 6. \tag{5.25}$$

Moreover this is the only condition because for all other choices the LHS of (5.24) is not invariant under M, the generator of a $U(1)$ subgroup of Lorentz group (see section 3.2.1 for the definition of M). Therefore the LHS is equal to zero. The solution is given by

$$\Lambda_{+++} = 6\delta(\lambda^+)\delta(\lambda_{12})\cdots\delta(\lambda_{45}). \tag{5.26}$$

It is possible to verify this object is indeed part of a representation of the Lorentz group. In order to do so one needs to check the Lorentz algebra holds when acting on Λ_{+++}. First note

$$(N_S)^a{}_b\Lambda_{+++} = N_{ab}\Lambda_{+++} = 0, \quad N\Lambda_{+++} = \frac{15}{4}\Lambda_{+++}, \tag{5.27}$$

N^{mn} denote the realisation of Lorentz generators M^{mn} in terms of pure spinors, see section 3.3 for the precise expressions. The nontrivial commutation relations that remain to be checked are

$$[N_{ab}, N^{cd}]\Lambda_{+++} = -\frac{1}{2}\delta^{[c}_{[a}N^{d]}{}_{b]}\Lambda_{+++} = -\frac{1}{5}\delta^c_{[a}\delta^d_{b]}N\Lambda_{+++} = -\frac{3}{4}\delta^c_{[a}\delta^d_{b]}\Lambda_{+++}, \tag{5.28}$$

$$[N^a_{\ b}, N^{cd}]\Lambda_{+++} = \frac{1}{2}\delta^{[c}_b N^{d]a}\Lambda_{+++}. \tag{5.29}$$

Because of the symmetric form of Λ_{+++} it suffices to check

$$[N_{12}, N^{12}]\Lambda_{+++} = -\frac{3}{4}\Lambda_{+++}, \tag{5.30}$$

$$[N_{12}, N^{13}]\Lambda_{+++} = 0, \tag{5.31}$$

$$[N^1{}_2, N^{23}]\Lambda_{+++} = -\frac{1}{2}N^{13}\Lambda_{+++}. \tag{5.32}$$

Let us start with the LHS of (5.30)

$$[N_{12}, N^{12}]\Lambda_{+++} = N_{12}N^{12}\Lambda_{+++} = N_{12}\left[\frac{1}{2}6\lambda^+\delta(\lambda^+)\delta'(\lambda_{12})\delta(\lambda_{13})\cdots\delta(\lambda_{45})\right] =$$

$$\frac{3}{2}(-w_+\lambda_{12} - \frac{1}{2}\frac{1}{\lambda^+}w^{ab}\lambda_{ab}\lambda_{12} + \frac{1}{\lambda^+}w^{ab}\lambda_{1a}\lambda_{2b})\left[\lambda^+\delta(\lambda^+)\delta'(\lambda_{12})\delta(\lambda_{13})\cdots\delta(\lambda_{45})\right] =$$

$$= (0 - \frac{9}{4} + \frac{3}{2})\Lambda_{+++} = -\frac{3}{4}\Lambda_{+++}, \tag{5.33}$$

Note that N_{12} does not contain factors of $(\lambda_{12})^2$ (possible such factors cancel out). This is useful when acting with N_{12} in this second line. In going from the second to the last line $x\delta'(x) = -\delta(x)$ was used twice. The other two commutators, (5.31) and (5.32), follow along the same lines.

It is instructive to compute the next two levels (distinguished by N charge) of the components of $\Lambda_{\alpha\beta\gamma}$. For the components on the second ($N = \frac{11}{4}$) level consider

$$N^{a_1a_2}\Lambda_{+++} = -\frac{1}{2}\Lambda^{a_1a_2}{}_{++} - \frac{1}{2}\Lambda_{+}{}^{a_1a_2}{}_{+} - \frac{1}{2}\Lambda_{++}{}^{a_1a_2} = -\frac{3}{2}\Lambda^{a_1a_2}{}_{++} \Rightarrow \tag{5.34}$$

$$\Lambda^{a_1a_2}{}_{++} = -\frac{2}{3}N^{a_1a_2}\Lambda_{+++}.$$

The factor of $-\frac{1}{2}$ is consistent with $N^{ab}w_+ = -\frac{1}{2}w^{ab}$. Going to the next level ($N = \frac{7}{4}$)

$$N^{b_1b_2}\Lambda^{a_1a_2}{}_{++} = -\frac{1}{2}\epsilon^{a_1a_2b_1b_2e}\Lambda_{e++} - \frac{1}{2}\Lambda^{a_1a_2b_1b_2}{}_{+} - \frac{1}{2}\Lambda^{a_1a_2}{}_{+}{}^{b_1b_2} = \tag{5.35}$$

$$-\frac{1}{2}\epsilon^{a_1a_2b_1b_2e}\Lambda_{e++} - \Lambda^{a_1a_2b_1b_2}{}_{+}.$$

This seems to leave freedom to define one of the two components, which would indeed be true if $\Lambda_{\alpha\beta\gamma}$ was just a symmetric rank three tensor and nothing more. However $\Lambda_{\alpha\beta\gamma}$ is gamma matrix traceless,

$$\gamma_m^{\alpha\beta}\Lambda_{\alpha\beta\gamma} = 0. \tag{5.36}$$

This imposes one additional condition that relates components of equal N charge to each other. Consequently all components of $\Lambda_{\alpha\beta\gamma}$ are uniquely fixed in terms of Λ_{+++}. Note that this is consistent with the discussion under (3.134), where Lorentz invariance arguments were used to come to the same conclusion.

Decoupling of Q_S exact states

The new insertion $\Lambda_{\alpha\beta\gamma}$ was motivated by manifest Lorentz invariance, but it also results in a prescription in which Q_S exact states decouple. Indeed, the tree-level amplitude with one Q_S exact state,

$$\mathcal{B} = \langle Q_S\Omega(z_1)V_2(z_2)V_3(z_3)\prod_{i=4}^{N}\int dz_iU(z_i)(\epsilon T)^{\delta_1\delta_2\delta_3}_{\beta_1\cdots\beta_{11}}\theta^{\beta_1}\cdots\theta^{\beta_{11}}(\infty)\Lambda_{\delta_1\delta_2\delta_3}(\infty)\rangle, \tag{5.37}$$

can be written in the following form:

$$\mathcal{B} = \langle\lambda^\alpha(z_2)\lambda^\beta(z_3)f_{\alpha\beta}(\theta)Q_S((\epsilon T)^{\delta_1\delta_2\delta_3}_{\beta_1\cdots\beta_{11}}\theta^{\beta_1}\cdots\theta^{\beta_{11}}\Lambda_{\delta_1\delta_2\delta_3})\rangle = \tag{5.38}$$

$$\langle\lambda^\alpha(z_2)\lambda^\beta(z_3)f_{\alpha\beta}(\theta)(\epsilon T)^{\delta_1\delta_2\delta_3}_{\beta_1\cdots\beta_{11}}\lambda^{\beta_1}\theta^{\beta_2}\cdots\theta^{\beta_{11}}\Lambda_{\delta_1\delta_2\delta_3}\rangle. \tag{5.39}$$

After using the OPE's to integrate out the nonzero modes one gets:

$$\mathcal{B} = \int d^{16}\theta[d\lambda]\lambda^\alpha\lambda^\beta f_{\alpha\beta}(\theta)(\epsilon T)^{\delta_1\delta_2\delta_3}_{\beta_1\cdots\beta_{11}}\lambda^{\beta_1}\theta^{\beta_2}\cdots\theta^{\beta_{11}}\Lambda_{\delta_1\delta_2\delta_3} = \tag{5.40}$$

$$\int d^{16}\theta f_{\alpha\beta}(\theta)(\epsilon T)^{\alpha\beta\beta_1}_{\beta_1\cdots\beta_{11}}\theta^{\beta_2}\cdots\theta^{\beta_{11}} = 0.$$

The last line vanishes because all traces of (ϵT) vanish (cf. (5.21)).

5.1.3 Global issues

The computations in section 5.1.1 showed that not all Q_S exact states decouple. From this result it is tempting to conclude that the PCOs are not Q_S closed. This is true as will be shown in section 5.3, but one cannot conclude it just yet. For the computations above involve integrations over the space of pure spinors, which is a manifold that cannot be covered by one coordinate patch. Therefore the computations in section 5.1.1 can best be viewed as evidence for the need of a globally well defined PCO. Note however that when one integrates over C, Lorentz invariance can be used and consequently any possible ambiguity goes away. Alternatively one can use Λ, globally defined in (5.24), which will be done in the one-loop computations in the next section.

5.2 One loop

In this section one-loop amplitudes with one unphysical state are considered both in the prescription with an integral over B and without. Let us first consider the case in which there is no B integration. All amplitudes, including those with an unphysical state, can be evaluated by first integrating out the nonzero modes. One is then left with a certain zero mode integral. At tree level one could show that these integrals vanish after the λ integration is performed, cf. (5.40). This section contains the corresponding one-loop computation. The result is that the zero mode integrals do not vanish after the λ, N integrations.

The analysis of amplitudes with an unphysical state when one includes an integral over B is analogous to the tree-level case. After one has integrated out the nonzero modes the zero mode integral over λ and N can be performed by Lorentz invariance. Recall that decoupling of unphysical states at tree level followed from the vanishing of the trace of ϵT cf. (5.40). This ϵT showed up in the λ zero mode integral (5.15). The analogous one-loop zero mode integral can be evaluated to give the one-loop analog of ϵT. Moreover one-loop amplitudes with an unphysical state are proportional to the trace of this one-loop invariant tensor. However this trace does not vanish. Therefore the question whether Q_S exact states decouple remains unanswered in this section. The computation including the B integral does show that the PCOs are not Q_S closed. In section 5.4 it will be shown using a different argument that unphysical states decouple to all orders, when one integrates over B and C.

5.2.1 No B integration

A one-loop amplitude with one unphysical state is given by

$$\mathcal{B}^{(N)} = \langle Q_S \Omega_1(z_1) \prod_{i=2}^{N} \int dz_i U_i(z_i) \int du \mu(u) \tilde{b}_{B^1}(u,w)(\lambda B^2 d)(y) \cdots (\lambda B^{10} d)(y)$$

$$(\lambda d)(y)\delta(B^1N(y))\cdots\delta(B^{10}N(y))\delta(J(y))\Lambda_{\delta_1\delta_2\delta_3}(y)(\epsilon T)^{\delta_1\delta_2\delta_3}_{\beta_1\cdots\beta_{11}}\theta^{\beta_1}(y)\cdots\theta^{\beta_{11}}(y)\rangle, \tag{5.41}$$

where $\lambda Bd = B_{mn}\lambda\gamma^{mn}d$. Note that the Y_C insertions have been replaced by the Lorentz covariant insertion, $\Lambda_{\alpha\beta\gamma}$, as in the tree-level computation. This is equivalent with inserting Y_C and integrating over C. On the torus one cannot insert the PCOs such that all their OPE's would vanish. They are inserted at some arbitrary point y. For later convenience $\tilde{b}$ is inserted at a different point, w.

The next step is integrating Q_S by parts. When Q_S acts on $\tilde{b}$ one gets a total derivative in moduli space, as usual. If this total derivative is non-vanishing the theory has a BRST anomaly. These total derivative terms will be suppressed below because they are not important for our discussion. The terms that come from Q_S hitting a picture raising operator, Z_B, vanish since the Q_S variation vanishes without subtleties, cf. (2.32). The vertex operators are also Q_S closed. The only non-vanishing terms come from Q_S hitting a θ. This results in a λ^{β_1} contracted with $\Lambda_{\alpha\beta\gamma}(\epsilon T)^{\alpha\beta\gamma}_{\beta_1\cdots\beta_{11}}$, very similar to tree-level amplitudes with an unphysical state. However the one-loop pure spinor zero mode integration also involves N^{mn}. As will be shown

$$\lambda^{\beta_1}\Lambda_{\alpha\beta\gamma}(\epsilon T)^{\alpha\beta\gamma}_{\beta_1\cdots\beta_{11}} \tag{5.42}$$

does not vanish after the one-loop pure spinor zero mode integrals have been performed.

After integrating Q_S by parts the amplitude (5.41) becomes

$$\mathcal{B}^{(N)} = \langle\Omega_1(z_1)\prod_{i=2}^{N}\int dz_iU_i(z_i)\int d\mu\mu(u)\tilde{b}_{B^1}(u,w)(\lambda B^2d)(y)\cdots(\lambda B^{10}d)(y)(\lambda d)(y) \tag{5.43}$$
$$\delta(B^2N(y))\cdots\delta(B^{10}N(y))\delta(J(y))\Lambda_{\delta_1\delta_2\delta_3}(y)(\epsilon T)^{\delta_1\delta_2\delta_3}_{\beta_1\cdots\beta_{11}}(y)\lambda^{\beta_1}(y)\theta^{\beta_2}(y)\cdots\theta^{\beta_{11}}(y)\rangle.$$

In this subsection $\mathcal{B}^{(N)}$ will be evaluated without integrating over B. The particular choice of B used here is given by

$$(B^1)_{ab} = \delta^{[1}_a\delta^{2]}_b, \cdots, (B^{10})_{ab} = \delta^{[4}_a\delta^{5]}_b, \quad (B^I)^{ab} = (B^I)^a{}_b = 0. \tag{5.44}$$

The amplitude $\mathcal{B}^{(N)}$ can be evaluated by first integrating out the nonzero modes and then evaluating the zero mode integrals. The nonzero mode integration is a little tedious since there is quite a number of $N\lambda$ OPE's one has to consider. Therefore the nonzero mode integration is explained in detail after the subsection on the zero mode integrals. Once the nonzero mode integrals have been performed the amplitude $\mathcal{B}^{(N)}$ can be written as a sum of terms that are all proportional to a certain λ, N zero mode integral, $I_{\beta_2\cdots\beta_{11}}$. This integral contains (5.42). In the next section it will be shown $I_{\beta_2\cdots\beta_{11}}$ does not vanish. This non-vanishing does not prove that there exists a non-vanishing amplitude with a Q_S exact state, because there may be additional cancellations when one performs the remaining integrals. It does show however that the PCOs are not Q_S closed, i.e. (5.42) does not vanish when integrated against an arbitrary function.

Zero mode integral

After integrating out the nonzero modes, which is discussed in detail in the next subsection, one-loop amplitudes (5.43) can be written as a sum of terms that are proportional to the following zero mode integral,

$$I^{\alpha_1}_{\beta_1\beta_2\cdots\beta_{11}} \equiv \tag{5.45}$$

$$\int [d\lambda][dN]\lambda^{\alpha_1}(\lambda\gamma^{13}d)\cdots(\lambda\gamma^{45}d)(\lambda d)\delta(N^{12})\cdots\delta(N^{45})\delta(J)\Lambda_{\alpha\beta\gamma}(\epsilon T)^{\alpha\beta\gamma}_{\beta_1\cdots\beta_{11}}.$$

Note that there is one unintegrated vertex operator at one loop, which explains the presence of λ^{α_1}. The factors $\lambda\gamma^{ab}d\delta(N^{ab})$ originate from the picture raising operators Z_B and $\delta(N^{12})$ stems from the b ghost. If one of the states is an unphysical state, the amplitude can be written as a sum of terms proportional to the trace of $I^{\alpha_1}_{\beta_1\cdots\beta_{11}}$ which is called $I_{\beta_2\cdots\beta_{11}}$:

$$I_{\beta_2\cdots\beta_{11}} \equiv I^{\alpha_1}_{\alpha_1\beta_2\cdots\beta_{11}} = \tag{5.46}$$

$$\int [d\lambda][dN]\lambda^{\beta_1}(\lambda\gamma^{13}d)\cdots(\lambda\gamma^{45}d)(\lambda d)\delta(N^{12})\cdots\delta(N^{45})\delta(J)\Lambda_{\alpha\beta\gamma}(\epsilon T)^{\alpha\beta\gamma}_{\beta_1\cdots\beta_{11}}.$$

This integral is the one-loop analog of (5.40) (or (5.20)). Therefore this integral must vanish if the PCOs are Q_S closed. Note that, in spite of the notation, $I^{\alpha_1}_{\beta_1\beta_2\cdots\beta_{11}}$ is not manifestly Lorentz invariant. Whether it is Lorentz invariant remains to be seen. Let us proceed by evaluating $I_{\beta_2\cdots\beta_{11}}$.

After using expression (3.127) for $[dN]$ to evaluate the N integral in $I_{\beta_2\cdots\beta_{11}}$ one finds

$$I_{\beta_2\cdots\beta_{11}} = \int [d\lambda]\frac{1}{(\lambda^+)^8}\lambda^{\beta_1}(\lambda\gamma^{13}d)\cdots(\lambda\gamma^{45}d)(\lambda d)\Lambda_{\alpha\beta\gamma}(\epsilon T)^{\alpha\beta\gamma}_{\beta_1\cdots\beta_{11}}. \tag{5.47}$$

In this form it becomes apparent that the problems with factors of λ^+ in the denominator only increase at one loop. At this point one can only surmise this. To find a definitive answer one has to evaluate the λ integral. This can be done by expanding the integrand by powers of λ^+, using the explicit gamma matrix expression from section 3.2.4:

$$\frac{1}{(\lambda^+)^8}(\lambda\gamma^{13}d)\cdots(\lambda\gamma^{45}d)(\lambda d) = \tag{5.48}$$

$$(\lambda^+)^2 D_{12}d_+ + \frac{1}{2}\lambda^+\lambda_{a_1a_2}\left(D_{12}d^{a_1a_2} + \frac{1}{2}\epsilon^{aba_1a_2c}d_cD_{12ab}d_+\right)+$$

$$\frac{1}{8}\lambda_{a_1a_2}\lambda_{a_3a_4}\left(D_{12}\epsilon^{aa_1a_2a_3a_4}d_a + \epsilon^{aba_1a_2c}d_cD_{12ab}d^{a_3a_4}+\right.$$

$$\frac{1}{2}\epsilon^{aba_1a_2c}\epsilon^{dea_3a_4f}d_cd_fD_{12abde}d_+\Big) + \sum_{k=3}^{6}\frac{1}{(\lambda^+)^{k-2}}\lambda_{a_1a_2}\cdots\lambda_{a_{2k-1}a_{2k}}Y^{a_1\cdots a_{2k}}(d),$$

where

$$D = d^{12}\cdots d^{45}, \quad D_{a_1\cdots a_k} = \frac{\partial}{\partial d^{a_{k-1}a_k}}\cdots\frac{\partial}{\partial d^{a_1a_2}}D. \tag{5.49}$$

The $Y(d)$'s can be expressed in terms of the d's similar to the first three terms. Note that the minimal number of d_a's in $Y^{a_1\cdots a_{2k}}$ is $k-1$. This is the reason the series stops at $k=6$. The maximum number of d_a's in $Y^{a_1\cdots a_{2k}}$ is k. The λ integration of (5.48) can be evaluated term by term. $I_{\beta_2\cdots\beta_{11}}$ then becomes

$$I_{\beta_2\cdots\beta_{11}} = \sum_{k=0}^{6}(I_k)_{a_1\cdots a_{2k}\beta_2\cdots\beta_{11}}Y^{a_1\cdots a_{2k}}. \tag{5.50}$$

The integrals I_k are investigated order by order in the sequel of this subsection.

For $k=0,1,2$ one can use the definition of $\Lambda_{\alpha\beta\gamma}$, (5.24), and the fact that the invariant tensor (ϵT) is traceless, (5.21), to show the λ integrals vanish:

$$(I_0)_{\beta_2\cdots\beta_{11}} = \int[d\lambda]\lambda^{\beta_1}(\lambda^+)^2\Lambda_{\delta_1\delta_2\delta_3}(\epsilon T)^{\delta_1\delta_2\delta_3}_{\beta_1\cdots\beta_{11}} = (\epsilon T)^{++\beta_1}_{\beta_1\cdots\beta_{11}} = 0, \tag{5.51}$$

$$(I_1)_{a_1a_2\beta_2\cdots\beta_{11}} = \int[d\lambda]\lambda^{\beta_1}\lambda^+\lambda_{a_1a_2}\Lambda_{\delta_1\delta_2\delta_3}(\epsilon T)^{\delta_1\delta_2\delta_3}_{\beta_1\cdots\beta_{11}} = (\epsilon T)^{+\beta_1}{}_{a_1a_2\ \beta_1\cdots\beta_{11}} = 0,$$

$$(I_2)_{a_1\cdots a_4\beta_2\cdots\beta_{11}} = \int[d\lambda]\lambda^{\beta_1}\lambda_{a_1a_2}\lambda_{a_3a_4}\Lambda_{\delta_1\delta_2\delta_3}(\epsilon T)^{\delta_1\delta_2\delta_3}_{\beta_1\cdots\beta_{11}} =$$

$$(\epsilon T)^{\beta_1}{}_{a_1a_2a_3a_4\ \beta_1\cdots\beta_{11}} = 0.$$

If $k>2$, however, there are also factors of λ^+ in the denominator. As shown in appendix A.1 the λ integrals do not vanish anymore. For example consider the integral I_3. By M charge conservation all components of I_3 vanish except when the indices are chosen to be

$$\beta_2,\ldots,\beta_{11} = +, b_1b_2,\ldots,b_9b_{10}, c_1, c_2, c_3, c_4 \tag{5.52}$$

or

$$\beta_2,\ldots,\beta_{11} = b_1b_2,\ldots,b_{13}b_{14}, c_1, c_2, c_3.$$

This is explained in detail in the first part of appendix A. Let us explicitly compute I_3 for the first choice of indices. Since $\mathrm{Sym}^3\overline{\mathbf{10}}\otimes\mathrm{Asym}^5\mathbf{10}\otimes\mathrm{Asym}^4\mathbf{5}$ contains one scalar, one finds

$$(I_3)_{a_1\cdots a_6+c_1c_2c_3c_4}{}^{b_1b_2\cdots b_9b_{10}} = \tag{5.53}$$

$$\int[d\lambda]\frac{1}{\lambda^+}\lambda^{\beta_1}\lambda_{a_1a_2}\cdots\lambda_{a_5a_6}\Lambda_{\alpha\beta\gamma}(\epsilon T)^{\alpha\beta\gamma}{}_{\beta_1+c_1c_2c_3c_4}{}^{b_1\cdots b_{10}} =$$

$$c_1\epsilon_{a_1a_2a_3a_4b_{14}}\epsilon_{b_{16}a_5a_6b_{11}b_{12}}(\epsilon_{10})^{b_1\cdots b_{20}}\epsilon_{c_1c_2c_3c_4b_{17}}\epsilon_{b_{13}b_{15}b_{18}b_{19}b_{20}} + 2 \text{ perms},$$

where $(\epsilon_{10})^{b_1\cdots b_{20}}$ is antisymmetric under both $b_{2i-1} \leftrightarrow b_{2i}$ and $b_{2i-1}b_{2i} \leftrightarrow b_{2j-1}b_{2j}$ and $(\epsilon_{10})^{12131415232425343545} = 1$. The two permutations add terms to make the RHS symmetric under $a_{2i-1}a_{2i} \leftrightarrow a_{2j-1}a_{2j}$. The constant c_1 is computed in appendix A.2 and is given by

$$c_1 = \frac{129}{2}. \tag{5.54}$$

The integral I_4 can be computed similarly, but this computation will not be presented here. The next integral is I_5. The only choice of $\beta_2, \ldots, \beta_{11}$ that leads to a nonzero answer for I_5 is

$$(I_5)_{a_1\cdots a_{10}}{}^{b_3\cdots b_{12}}{}_{12345} = \int [d\lambda]\frac{1}{(\lambda^+)^3}\lambda^{\beta_1}\lambda_{a_1a_2}\cdots\lambda_{a_9a_{10}}\Lambda_{\delta_1\delta_2\delta_3}(\epsilon T)^{\delta_1\delta_2\delta_3}{}_{\beta_1}{}^{b_3\cdots b_{12}}{}_{12345} =$$

$$-\frac{2}{5}\epsilon_{b_{13}a_1a_2a_3a_4}\epsilon_{b_{15}a_5a_6a_7a_8}\epsilon_{b_{17}a_9a_{10}b_1b_2}(\epsilon_{10})^{b_1\cdots b_{20}}\epsilon_{b_{14}b_{16}b_{18}b_{19}b_{20}} + 14 \text{ perms}. \tag{5.55}$$

The details are given in appendix A.2. Finally I_6 can be evaluated as:

$$(I_6)_{a_1\cdots a_{12}\beta_2\cdots\beta_{11}} = \int [d\lambda]\frac{1}{(\lambda^+)^4}\lambda^{\beta_1}\lambda_{a_1a_2}\cdots\lambda_{a_{11}a_{12}}\Lambda_{\alpha\beta\gamma}(\epsilon T)^{\alpha\beta\gamma}_{\beta_1\cdots\beta_{11}} = \tag{5.56}$$

$$\epsilon_{b_1a_1a_2a_3a_4}\epsilon_{b_2a_5a_6a_7a_8}\epsilon_{b_3a_9a_{10}a_{11}a_{12}}(\epsilon T)^{b_1b_2b_3}_{+\beta_2\cdots\beta_{11}} + \text{permutations} = 0.$$

This vanished because $(\epsilon T)^{b_1b_2b_3}_{+\beta_2\cdots\beta_{11}} = 0$ and that follows from the M charge conservation rule for invariant tensors. In other words it is not possible to choose $\beta_2, \ldots, \beta_{11}$ such that the total M charge of the components is zero (cf. equation (3.54)). This concludes the computation of the pure spinor zero mode integrals that appear at one loop. It has been shown that the Q_S variation of the PCO as given in (5.42) does not vanish after the integration over the pure spinor sector in a typical one-loop zero mode integral. Therefore the PCOs are not Q_S closed.

Nonzero mode integration

It remains to demonstrate that all one-loop amplitudes with an unphysical state can be written as a sum of terms proportional to $I_{\beta_2\cdots\beta_{11}}$. After this proof the argument will be modified to prove that $\mathcal{A}^{(N)}$ can be written as a sum of terms proportional to $I^{\alpha}_{\beta_1\cdots\beta_{11}}$. In general the amplitude, $\mathcal{B}^{(N)}$, becomes a sum of terms of the form

$$\mathcal{B}^{(N)}_{i_1\cdots i_k} = \int [\mathcal{D}\lambda][\mathcal{D}N][\mathcal{D}d][\mathcal{D}\theta](\prod_{i=2}^{N}\int dz_i) f_{m_1n_1\cdots m_kn_k}(z_1, \ldots, z_N) \tag{5.57}$$

$$N^{m_1n_1}(z_{i_1})\cdots N^{m_kn_k}(z_{i_k})(\lambda\gamma^{13}d)(y)\cdots(\lambda\gamma^{45}d)(y)(\lambda d)(y)\lambda^{\beta_1}(y)\Lambda_{\alpha\beta\gamma}(y)(\epsilon T)^{\alpha\beta\gamma}_{\beta_1\cdots\beta_{11}}$$

$$\theta^{\beta_2}(y)\cdots\theta^{\beta_{11}}(y)\int du\mu(u)\tilde{b}_{B^1}(u,w)\delta(N^{13}(y))\cdots\delta(N^{45}(y))\delta(J(y))e^{-S},$$

where the indices in the PCOs are $SU(5)$ indices, $i_j \in \{2,\ldots,N\}$ and $f_{m_1\cdots n_k}$ does not contain any λ's or w's. The number k indicates how many vertex operators provide an N^{mn}. The functional integrals over λ and N can be evaluated by performing the OPE's to remove all fields of nonzero weight. Then one replaces the fields by their zero modes and performs the integration over these modes. In order to perform the OPE between N^{mn} and $\delta(BN)$ one has to Taylor expand $\delta(BN)$, as discussed in [22],

$$\delta((BN(y)) = \delta(BN_0\omega(y) + B\hat{N}(y)) = \tag{5.58}$$
$$\delta(BN_0\omega(y)) + (B\hat{N}(y))\delta'(BN_0\omega(y)) + \frac{1}{2}(B\hat{N}(y))^2\delta''(BN_0\omega(y)) + \cdots,$$

where $\hat{N}$ denotes N after omission of the zero mode. The holomorphic one form $\omega(y)$ is constant on the torus:

$$\omega(y) = \frac{1}{4\pi^2\tau_2}, \tag{5.59}$$

where τ_2 is the imaginary part of the modulus τ. The b ghost also contains N^{mn}'s which have to be taken into account if one is removing all fields of nonzero weight. Let us start with the first term, the local b ghost, $b_B(u)$. The second term of $\tilde{b}(u,y)$, with the integration in it, will be dealt with later. After replacing $\tilde{b}(u,y)$ by $b(u)$ in the amplitude, $\mathcal{B}^{(N)}_{i_1\cdots i_k}$, becomes a sum over n, which counts the number of N^{mn}'s the local b ghost provides, of the following objects:

$$\mathcal{B}^{(N)}_{i_1\cdots i_k,n} = \int [\mathcal{D}\lambda][\mathcal{D}N][\mathcal{D}d][\mathcal{D}\theta](\prod_{i=2}^{N}\int dz_i)\int du\mu(u)\sum_{j=0}^{3} f_{jm_1n_1\cdots m_{k+n}n_{k+n}}(z,u,w)$$
$$N^{m_1n_1}(z_{i_1})\cdots N^{m_kn_k}(z_{i_k})N^{m_{k+1}n_{k+1}}(w)\cdots N^{m_{k+n}n_{k+n}}(w)(\lambda\gamma^{13}d)(y)\cdots(\lambda\gamma^{45}d)(y)$$
$$(\lambda d)(y)\lambda^{\beta_1}(y)(\epsilon T)^{\alpha\beta\gamma}_{\beta_1\cdots\beta_{11}}\Lambda_{\alpha\beta\gamma}(y)\theta^{\beta_2}(y)\cdots\theta^{\beta_{11}}(y)$$
$$\delta^{(j)}(N^{12}(w))\delta(N^{13}(y))\cdots\delta(N^{45}(y))\delta(J(y))e^{-S}, \tag{5.60}$$

where $\delta^{(j)}$ denotes the jth derivative of the delta function and the sum runs from zero to three because b does not contain $\delta^{(4)}(B^1N)$ or higher derivatives.

The product of the eleven delta functions, including the one from b, becomes a sum of products of eleven $\delta^{(j)}(B^IN_0)$ after the Taylor expansion. Let us start with the first term in this sum, i.e. the one without $\hat{N}$'s and no derivatives on the delta functions. In this case the $N^{m_jn_j}(z_j)$'s from (5.57) have OPE's with themselves and with the λ's from the PCOs. Let us first concentrate on the term in which all N^{mn}'s get contracted with an explicit λ. That term is given by[4]

$$\mathcal{C}^{(N)}_{i_1\cdots i_k,n} = \int [d\lambda][dN][\mathcal{D}^{16}d][\mathcal{D}^{16}\theta]\left[\prod_{i=2}^{N}\int dz_i\right]\int du f_{m_1n_1\cdots m_{k+n}n_{k+n}}(z_1,\cdots,z_N,u)$$

[4]Since the distinction between worldsheet fields and their zero modes plays a central role in the argument, zero modes are denoted in an explicit way, unlike in other parts of this work.

$$\left[\prod_{l=1}^{k} F(z_{i_l}, y)\right] F(w,y)^n \mathcal{N}^{m_1 n_1} \cdots \mathcal{N}^{m_{k+n} n_{k+n}} \lambda_0^{\beta_1} (\lambda_0 \gamma^{13} d(y)) \cdots (\lambda_0 \gamma^{45} d(y))$$

$$(\lambda_0 d(y))(\Lambda_0)_{\alpha\beta\gamma}(\epsilon T)^{\alpha\beta\gamma}_{\beta_1\cdots\beta_{11}} \theta^{\beta_2}(y)\cdots\theta^{\beta_{11}}(y)\delta(N_0^{12})\cdots\delta(N_0^{45})\delta(J_0)e^{S_{p\theta}}, \quad (5.61)$$

where

$$F(z,y) = \partial_z \log E(z,y) \quad (5.62)$$

and $E(z,y)$ is the holomorphic prime form, which goes like $z-y$ when $z \to y$ [3, 18]. $\mathcal{N}^{mn}$ are abstract Lorentz generators for the λ, w sector and they act to the right. They should not be thought of as containing (zero) modes of the λ or w worldsheet fields. The $\mathcal{N}^{mn}$ merely multiply every index on a λ or w they hit by a two form gamma matrix. Up to now only contractions between N^{mn} and the explicit λ's have been considered, but if two or more N^{mn}'s contract with each other in $\mathcal{B}^{(N)}_{i_1\cdots i_k,n}$ one gets a term of the form $\mathcal{C}^{(N)}_{i_1\cdots i_l,m}$, with $l+m<k+n$, where the poles in $z_i - z_j$ are included in the unspecified function f.

The last step of our argument is showing all terms with derivatives on the delta functions can also be written as a sum of terms of the form $\mathcal{C}^{(N)}_{i_1\cdots i_k,n}$. To see this note that if a derivative acts on $\delta(N^{ab})$ one of the N^{mn} must provide this zero mode, otherwise the integral vanishes. This step just reduces the number of N^{mn}'s in $\mathcal{B}^{(N)}_{i_1\cdots i_k,n}$ that must be contracted, so in fact it becomes of the form $\mathcal{C}^{(N)}_{i_1\cdots i_l,m}$ where $k+n-l-m$ is the number derivatives acting on the delta functions. Since the zero mode measures $[d\lambda]$ and $[dN]$ are Lorentz invariant one can pull the $\mathcal{N}$ out of these integrals. This concludes the main part of the argument that a one-loop amplitude can be written as a sum of terms proportional to $I_{\beta_2\cdots\beta_{11}}$.

One still needs to consider the second term in $\tilde{b}(u,w)$. This was not included in the above discussion because it contains $\partial N^{mn}(v)$. This does not change the argument much, after the OPE's this part of the amplitude will also have the form of $\mathcal{C}^{(N)}_{i_1\cdots i_k,n}$ where the effect of the v derivative and the integral over v are included in f.

To see $\mathcal{A}^{(N)}$ can be written as a sum of terms proportional to $I^{\alpha_1}_{\beta_1\cdots\beta_{11}}$ one can use the above reasoning with a slight adjustment. This consists of replacing $\lambda^{\beta_1}(y)$ by $\lambda^{\alpha_1}(z_1)$ in (5.57) and adding an α_1 index to f. The only effect this has is the replacement of some $F(z_i,y)$ by $F(z_1,z_i)$ in (5.61), apart from the fact α_1 and β_1 are not contracted anymore.

Four point function

The one-loop four-point function with an unphysical state in the formulation without an integral over B vanishes. This should come as a surprise after the result of the previous section, where it was shown that the Q_S variation of the PCOs does not vanish. The vanishing of the amplitude is instead achieved after the integral over the d zero modes has been performed.

The one-loop four-point amplitude is an example of an amplitude in which only the zero modes contribute (cf. [22]). It turns out only three terms have enough factors of d_α and N^{mn} to give a non-vanishing answer. This will become clear in equation (5.64) below. Thus one can immediately replace all the fields in (5.41) by their zero modes:

$$\mathcal{B}^{(4)} = \int [d\lambda][dN]d^{16}dd^{16}\theta Q_S\Omega \prod_{i=2}^{4} U_i\tilde{b}_{B^1}(\lambda B^2 d)\cdots(\lambda B^{10}d)(\lambda d) \tag{5.63}$$

$$\delta(B^1N)\cdots\delta(B^{10}N)\delta(J)\Lambda_{\alpha\beta\gamma}(\epsilon T)^{\alpha\beta\gamma}_{\beta_1\cdots\beta_{11}}\theta^{\beta_1}\cdots\theta^{\beta_{11}}.$$

For the d integration to be non-vanishing there must be a total of sixteen d zero modes, therefore the only terms of b_{B^1} that contribute are the ones with four d's and there are only three such terms:

$$\begin{aligned}(b_B)|_{d^4} \;=\; & -\frac{1}{1536}\gamma^{\alpha\beta}_{mnp}(d\gamma^{mnp}d)(Bd)_\alpha(Bd)_\beta\delta'(BN) \\ & -\frac{1}{8}c_1{}^{\gamma\delta\alpha\rho}_{mn}N^{mn}d_\rho(Bd)_\alpha(Bd)_\beta(Bd)_\gamma\delta''(BN) \\ & -\frac{1}{16}c_4{}^{\delta\gamma\beta\alpha}_{mnpq}N^{mn}N^{pq}(Bd)_\alpha(Bd)_\beta(Bd)_\gamma(Bd)_\delta\delta'''(BN),\end{aligned} \tag{5.64}$$

where the invariant tensors c_1 and c_4 can be read off from (2.39)-(2.42) and (3.139)-(3.148). Note the N integration will only be non-vanishing if the fourth vertex operator provides an N^{mn} zero mode. Moreover there are no terms in the b ghost with three d's and no derivatives on $\delta(BN)$. Such terms could have contributed here. The three terms above turn out to all be proportional to (for $B_{ab} = \delta^1_{[a}\delta^2_{b]}, B^a{}_b = B^{ab} = 0$)

$$d^{12}d_3d_4d_5\delta'(N^{12}). \tag{5.65}$$

For the first term this follows from direct computation using the gamma matrices as listed in section 3.2.4. Actually, one could have predicted the fact that three of the four d_α's are d_a's and one is a d^{ab}, by looking at the M charge of the full term. $\delta'(N^{12})$ has M charge two and since $\gamma^{\alpha\beta}_{mnp}(d\gamma^{mnp}d)(Bd)_\alpha(Bd)_\beta\delta'(BN)$ has M charge zero, the d part must have M charge minus two. The only way four d's can give M charge minus two is when three of them are a d_a (M charge $-\frac{3}{4}$) and the fourth is a d^{ab} (M charge $\frac{1}{4}$).

The second term can be reduced as follows:

$$(c_1)^{\gamma\beta\alpha\rho}_{mn}N^{mn}d_\rho(Bd)_\alpha(Bd)_\beta(Bd)_\gamma\delta''(BN) = \tag{5.66}$$

$$(c_1)_{12\ a_1\cdots a_8}d^{a_7a_8}\frac{1}{2}\epsilon^{a_1a_212a}d_a\frac{1}{2}\epsilon^{a_3a_412b}d_b\frac{1}{2}\epsilon^{a_5a_612c}d_c\delta'(N^{12}),$$

where the M charge conservation property of invariant tensors was used together with $(Bd)_a = 0$. After observing that $(c_1)_{aba_1\cdots a_8}$ is an $SU(5)$ invariant tensor that is antisymmetric in the middle three pairs of indices (a_1a_2, a_3a_4, a_5a_6) and there is only one invariant tensor with these symmetries [40], namely $(\epsilon_{aba_1a_2[a_3}\epsilon_{a_4]a_5a_6a_7a_8}+$ 5 perms), one finds that the second term in the b ghost is proportional to

$$(c_1)_{12343545a_7a_8}d^{a_7a_8}d_3d_4d_5\delta'(N^{12}) = d^{12}d_3d_4d_5\delta'(N^{12}). \tag{5.67}$$

The same logic can be applied to the third term although this case is slightly simpler. $\alpha, \beta, \gamma, \delta$ has to be $+, ab, cd, ef$ and since $(Bd)_+ = d^{12}$ one automatically gets this factor.

The third integrated vertex operator must provide an N^{12} zero mode. It then follows that $\mathcal{B}^{(4)}$ is proportional to $I_{\beta_2\cdots\beta_{11}}$. This integral can be written as a sum over k just as in (5.48). In this sum the $k=0,1,2,6$ terms vanish because of the λ integration and the $k=4,5$ terms vanish due to the d integration (note that b_{B^1} contains three d_a's and Y_4, Y_5 contain at least three d_a's). The $k=3$ term is given by

$$(b_{B^1})|_{d^4}(I_3)_{a_1\cdots a_6\beta_2\cdots\beta_{11}}(Y_3)^{a_1\cdots a_6} = \frac{1}{32}d^{12}d_3d_4d_5\Bigg(\epsilon^{aba_1a_2c}d_cD_{12ab}\epsilon^{da_3a_4a_5a_6}d_d+ \tag{5.68}$$

$$\epsilon^{aba_1a_2c}\epsilon^{dea_3a_4f}d_cd_fD_{12abde}d^{a_5a_6} + \frac{1}{2}\epsilon^{aba_1a_2c}\epsilon^{dea_3a_4f}\epsilon^{gha_5a_6j}d_cd_fd_jD_{12abdegh}d_+\Bigg)$$

$$\int[d\lambda]\frac{1}{\lambda^+}\lambda^{\beta_1}\lambda_{a_1a_2}\lambda_{a_3a_4}\lambda_{a_5a_6}\Lambda_{\delta_1\delta_2\delta_3}(\epsilon T)^{\delta_1\delta_2\delta_3}_{\beta_1\cdots\beta_{11}} =$$

$$-\frac{1}{4}d^{12}d_3d_4d_5\epsilon^{aba_1a_2c}d_cd_dD_{12ab}\int[d\lambda]\lambda^{\beta_1}\lambda^d\lambda_{a_5a_6}\Lambda_{\delta_1\delta_2\delta_3}(\epsilon T)^{\delta_1\delta_2\delta_3}_{\beta_1\cdots\beta_{11}}+$$

$$-d^{12}d_3d_4d_5\epsilon^{aba_1a_2c}d_cd_fD_{12ab}\int[d\lambda]\lambda^{\beta_1}\lambda_{a_1a_2}\lambda^f\Lambda_{\delta_1\delta_2\delta_3}(\epsilon T)^{\delta_1\delta_2\delta_3}_{\beta_1\cdots\beta_{11}} = 0,$$

where the following identity was used

$$D_{12abcd}d^{ef} = -\delta_c^{[e}\delta_d^{f]}D_{12ab} - \delta_1^{[e}\delta_2^{f]}D_{abcd} - \delta_a^{[e}\delta_b^{f]}D_{cd12} \tag{5.69}$$

and the integral vanishes because ϵT is traceless.

Thus, for the four-point one-loop amplitudes with a Q_S exact state the terms that do not vanish after the λ, N integral now vanish because they contain a square of fermionic quantity, namely $d_\alpha d_\alpha$ (no sum). Decoupling of unphysical states in higher point function is much more tedious to check since the nonzero mode integrations are non-trivial and the lack of manifest Lorentz invariance.

5.2.2 Including B integration

At tree level decoupling of unphysical states was restored after integrating over the constant spinors C. In this section manifest Lorentz invariance for one-loop amplitudes is restored by including the B integration. Whether this leads to decoupling of unphysical is the subject of this section. Similar to the tree-level case one can show that all amplitudes are proportional to a certain invariant tensor (at tree level this was (ϵT)) and amplitudes with Q_S exact states are proportional to the trace of this invariant tensor. However, at one loop the trace of this tensor does not vanish.

Following the same steps as in the previous subsection (section 5.3 contains details of these steps), one can show that all amplitudes can be written as a sum of terms proportional to the following zero mode integral

$$X^{\alpha_1\cdots\alpha_{11}}_{\beta_1\cdots\beta_{11}m_1n_1\cdots m_{10}n_{10}} \equiv \int [dB][dC][d\lambda][dN]\lambda^{\alpha_1}\cdots\lambda^{\alpha_{11}} \tag{5.70}$$

$$B^1_{m_1n_1}\cdots B^{10}_{m_{10}n_{10}}C^1_{\beta_1}\cdots C^{11}_{\beta_{11}}\delta(C^1\lambda)\cdots\delta(C^{11}\lambda)\delta(B^1N)\cdots\delta(B^{10}N)\delta(J).$$

Proportional here means in the sense of tensor multiplication: in the terms that appear after contractions, the tensor X is multiplied by gamma matrices. Evaluating the integrals in (5.70) is much easier than one might have anticipated, because X must be an invariant tensor, that is symmetric and gamma matrix traceless in the α's, antisymmetric in the β's and antisymmetric in both $m_i \leftrightarrow n_i$ and $m_in_i \leftrightarrow m_jn_j$. To find out how many independent invariant tensors with these properties exist, one has to compute the number of scalars in the relevant tensor product, which is one (see also section 3.2.3). The relevant invariant tensor has already appeared in the one-loop prescription in chapter 2:

$$(\epsilon TR)^{\alpha_1\cdots\alpha_{11}}_{\beta_1\cdots\beta_{11}m_1n_1\cdots m_{10}n_{10}} \equiv (\epsilon T)^{((\alpha_1\alpha_2\alpha_3}_{\beta_1\cdots\beta_{11}} R^{\alpha_4\cdots\alpha_{11}))}_{m_1n_1\cdots m_{10}n_{10}}, \tag{5.71}$$

where the double brackets denote gamma matrix traceless, cf. section 3.5. Lorentz invariance has completely fixed X, there is no freedom remaining.

Starting from a correlator with an unphysical state and integrating Q_S by parts, it will hit a θ from a PCO (where the total derivative in moduli space obtained when Q_S acts on $\tilde{b}$ is again suppressed, this derivative does not play a role here). This means all amplitudes with an unphysical state can be written as a sum of terms proportional to the trace of (ϵTR):

$$\int [dB][dC][d\lambda][dN]\lambda^{\alpha_2}\cdots\lambda^{\alpha_{11}}B^1_{m_1n_1}\cdots B^{10}_{m_{10}n_{10}}\lambda^{\beta_1}C^1_{\beta_1}C^2_{\beta_2}\cdots C^{11}_{\beta_{11}}$$

$$\delta(C^1\lambda)\cdots\delta(C^{11}\lambda)\delta(B^1N)\cdots\delta(B^{10}N)\delta(J) = (\epsilon TR)^{\alpha_1\cdots\alpha_{11}}_{\alpha_1\beta_2\cdots\beta_{11}m_1n_1\cdots m_{10}n_{10}}. \tag{5.72}$$

There are two independent invariant tensors with indices and symmetries of the trace of (ϵTR), so one expects a non-vanishing trace. Indeed, it is proved in section 5.2.3 that this trace does not vanish, which provides another proof for the fact that the PCO is not Q_S closed. The non-vanishing of the trace implies the proof of decoupling of unphysical states at tree level does not generalise to one loop and one needs a new argument. Such a new argument is presented in section 5.4, where it is shown that unphysical states decouple to all loop order.

Comparison to non-minimal formalism

In this subsection a brief comparison with the non-minimal formalism [25] is made. In this case all insertions are Q_S closed and decoupling of unphysical states follows straightforwardly.

In the non-minimal formalism the PCOs are replaced by

$$\mathcal{N} = e^{-(\lambda\bar{\lambda}+r\theta+\frac{1}{2}N_{mn}\bar{N}^{mn}+\frac{1}{4}S_{mn}\lambda\gamma^{mn}d+J\bar{J}+\frac{1}{4}S\lambda d)}. \tag{5.73}$$

This is invariant under Q_S:

$$Q_S\mathcal{N} = (\lambda r - \lambda r + \bar{N}^{mn}\frac{1}{2}\lambda\gamma_{mn}d - \bar{N}^{mn}\frac{1}{2}\lambda\gamma_{mn}d + \bar{J}(\lambda d) - \bar{J}(\lambda d))\mathcal{N} = 0. \tag{5.74}$$

Thus, all problematic terms of the minimal formalism are manifestly absent here and Q_S exact states decouple. In other words, these amplitudes vanish because two equal terms are subtracted.

5.2.3 Non-vanishing of the trace of (ϵTR)

In this subsection the trace Tr (ϵTR) of the tensor (ϵTR) is computed. To show that this trace does not vanish it is convenient to define a tensor Y and an operator $\hat{X}$:

$$\begin{aligned} Y_{m_1\cdots n_{10}} &\equiv \bar{\lambda}_{\alpha_4}\cdots\bar{\lambda}_{\alpha_{11}}R^{\alpha_4\cdots\alpha_{11}}_{m_1\cdots n_{10}}, & (5.75)\\ \hat{X} &\equiv \psi_{\beta_{12}}\cdots\psi_{\beta_{16}}\bar{\lambda}_{\alpha_1}\cdots\bar{\lambda}_{\alpha_3}T^{\beta_{12}\cdots\beta_{16},\alpha_1\alpha_2\alpha_3}\psi_\alpha\frac{\partial}{\partial\bar{\lambda}_\alpha}, & (5.76)\end{aligned}$$

where ψ_α is a fermionic Weyl spinor and $\bar{\lambda}_\alpha$ is a pure spinor of opposite chirality to λ^α. Note that, because $\bar{\lambda}_\alpha$ is a constrained spinor, $\partial/\partial\bar{\lambda}_\alpha$ is only defined up to a gauge transformation:

$$\delta\frac{\partial}{\partial\bar{\lambda}_\alpha} = A^m(\gamma_m\bar{\lambda})^\alpha. \tag{5.77}$$

The operator $\hat{X}$, however, is well defined, since it is gauge invariant. This follows from

$$\bar{\lambda}\gamma^q\psi\psi_{\beta_{12}}\cdots\psi_{\beta_{16}}\bar{\lambda}_{\alpha_1}\cdots\bar{\lambda}_{\alpha_3}T^{\beta_{12}\cdots\beta_{16},\alpha_1\alpha_2\alpha_3} = 0. \tag{5.78}$$

That can be shown be noting there are no scalars in $\text{Asym}^6\mathbf{16}'\otimes\mathbf{10}\otimes\text{Gam}^4\mathbf{16}'$, where Gam means the symmetric and gamma matrix traceless tensor product. Note one can use

$$\frac{\partial}{\partial\bar{\lambda}_\alpha}\bar{\lambda}_\beta = \delta^\alpha_\beta \tag{5.79}$$

when $\partial/\partial\bar{\lambda}_\alpha$ is part of a gauge invariant quantity, $S_\alpha(\partial/\partial\bar{\lambda}_\alpha)$, because

$$S_\alpha\frac{\partial}{\partial\bar{\lambda}_\alpha}\bar{\lambda}\gamma^m\bar{\lambda} = S\gamma^m\bar{\lambda} = 0, \tag{5.80}$$

the last equality is a consequence of gauge invariance.

The first step of the argument is showing that $\hat{X}Y \neq 0$. The second and last step is proving this implies the trace of (ϵTR) does not vanish. Consider the following component of $\hat{X}Y$ in a Lorentz frame in which the only nonzero component of $\bar{\lambda}$ is $\bar{\lambda}_+$:

$$\begin{aligned}\hat{X}Y_{a_1b_1a_2b_2\cdots a_{10}b_{10}} &= (\bar{\lambda}\gamma_m\psi)(\bar{\lambda}\gamma_n\psi)(\bar{\lambda}\gamma_p\psi)(\psi\gamma^{mnp}\psi) \\ &\Big[2(\psi\gamma_{a_1b_1a_2a_3a_4}\bar{\lambda})(\bar{\lambda}\gamma_{a_5b_5b_2a_6a_7}\bar{\lambda})(\bar{\lambda}\gamma_{a_8b_8b_3b_6a_9}\bar{\lambda})(\bar{\lambda}\gamma_{a_{10}b_{10}b_4b_7b_9}\bar{\lambda}) \\ &2(\bar{\lambda}\gamma_{a_1b_1a_2a_3a_4}\bar{\lambda})(\psi\gamma_{a_5b_5b_2a_6a_7}\bar{\lambda})(\bar{\lambda}\gamma_{a_8b_8b_3b_6a_9}\bar{\lambda})(\bar{\lambda}\gamma_{a_{10}b_{10}b_4b_7b_9}\bar{\lambda}) \\ &2(\bar{\lambda}\gamma_{a_1b_1a_2a_3a_4}\bar{\lambda})(\bar{\lambda}\gamma_{a_5b_5b_2a_6a_7}\bar{\lambda})(\psi\gamma_{a_8b_8b_3b_6a_9}\bar{\lambda})(\bar{\lambda}\gamma_{a_{10}b_{10}b_4b_7b_9}\bar{\lambda}) \\ &2(\bar{\lambda}\gamma_{a_1b_1a_2a_3a_4}\bar{\lambda})(\bar{\lambda}\gamma_{a_5b_5b_2a_6a_7}\bar{\lambda})(\bar{\lambda}\gamma_{a_8b_8b_3b_6a_9}\bar{\lambda})(\psi\gamma_{a_{10}b_{10}b_4b_7b_9}\bar{\lambda}) \\ &+\text{permutations}\Big],\end{aligned} \tag{5.81}$$

where the permutations make the RHS antisymmetric in $a_ib_i \leftrightarrow a_jb_j$. This reduces, up to an overall constant which is not zero[5], to

$$\hat{X}Y_{a_1b_1a_2b_2\cdots a_{10}b_{10}} = \epsilon^{c_1\cdots c_5}\psi_{c_1}\cdots\psi_{c_5}(\bar{\lambda}_+)^{10}\psi_+\gamma^{++}_{a_1b_1a_2a_3a_4}\gamma^{++}_{a_5b_5b_2a_6a_7} \tag{5.84}$$

$$\gamma^{++}_{a_8b_8b_3b_6a_9}\gamma^{++}_{a_{10}b_{10}b_4b_7b_9} + \text{permutations} = \epsilon^{c_1\cdots c_5}\psi_{c_1}\cdots\psi_{c_5}(\bar{\lambda}_+)^{10}\psi_+(\epsilon_{10})_{a_1\cdots b_{10}} \neq 0.$$

What remains is to show the non-vanishing of this tensor implies the non-vanishing of the trace of (ϵTR).

$$\hat{X}Y_{m_1n_1\cdots m_{10}n_{10}} = \epsilon^{\beta_1\cdots\beta_{16}}[(\epsilon T)^{((\alpha_1\alpha_2\alpha_3}_{\beta_1\cdots\beta_{11}}\psi_{\alpha_{11}}\psi_{\beta_{12}}\cdots\psi_{\beta_{16}}]R^{\alpha_4\cdots\alpha_{11}))}_{m_1n_1\cdots m_{10}n_{10}}\bar{\lambda}_{\alpha_1}\cdots\bar{\lambda}_{\alpha_{10}}. \tag{5.85}$$

For the term in the square brackets one can move the α_{11} to (ϵT) by using

$$\begin{aligned}0 = (\epsilon T)^{\alpha_1\alpha_2\alpha_3}_{[\beta_1\cdots\beta_{11}}\psi_{\beta_{12}}\cdots\psi_{\beta_{16}}\psi_{\alpha_{11}]} = \\ 6(\epsilon T)^{\alpha_1\alpha_2\alpha_3}_{[\beta_1\cdots\beta_{11}}\psi_{\beta_{12}}\cdots\psi_{\beta_{16}]}\psi_{\alpha_{11}} + 11(\epsilon T)^{\alpha_1\alpha_2\alpha_3}_{\alpha_{11}[\beta_1\cdots\beta_{10}}\psi_{\beta_{11}}\cdots\psi_{\beta_{16}]}.\end{aligned} \tag{5.86}$$

The first line is zero because it contains an antisymmetrisation of seventeen indices that only take sixteen values.

$$\hat{X}Y_{m_1n_1\cdots m_{10}n_{10}} = \epsilon^{\beta_1\cdots\beta_{16}}[(\epsilon T)^{((\alpha_1\alpha_2\alpha_3}_{\alpha_{11}\beta_1\cdots\beta_{10}}\psi_{\beta_{11}}\cdots\psi_{\beta_{16}}]R^{\alpha_4\cdots\alpha_{11}))}_{m_1n_1\cdots m_{10}n_{10}}\bar{\lambda}_{\alpha_1}\cdots\bar{\lambda}_{\alpha_{10}}. \tag{5.87}$$

Since $(\epsilon TR)^{\alpha_1\cdots\alpha_{11}}_{\alpha_{11}\beta_2\cdots\beta_{11}m_1n_1\cdots m_{10}n_{10}}$ is fully antisymmetric in $\beta_2\cdots\beta_{11}$ and symmetric and gamma matrix traceless in $\alpha_1\cdots\alpha_{10}$, one can conclude from the non-vanishing of $\hat{X}Y$ that

$$(\epsilon T)^{((\alpha_1\alpha_2\alpha_3}_{\alpha_{11}\beta_1\cdots\beta_{10}}R^{\alpha_4\cdots\alpha_{11}))}_{m_1n_1\cdots m_{10}n_{10}} \neq 0. \tag{5.88}$$

[5]Constants were omitted in the following two relations:

$$(\bar{\lambda}\gamma_m\psi)(\bar{\lambda}\gamma_n\psi)(\bar{\lambda}\gamma_p\psi)(\psi\gamma^{mnp}\psi) \propto \epsilon^{c_1\cdots c_5}\psi_{c_1}\cdots\psi_{c_5}(\bar{\lambda}_+)^3, \tag{5.82}$$

$$(\gamma^{++}_{a_1b_1a_2a_3a_4}\gamma^{++}_{a_5b_5b_2a_6a_7}\gamma^{++}_{a_8b_8b_3b_6a_9}\gamma^{++}_{a_{10}b_{10}b_4b_7b_9} + \text{permutations}) \propto (\epsilon_{10})_{a_1b_1\cdots a_{10}b_{10}}. \tag{5.83}$$

5.3 No-go theorem for Q_S closed, Lorentz invariant PCOs

In the previous section it was proved that the Lorentz invariant PCO was not Q_S closed. A logical next step is to give a modified prescription in which the PCO is Q_S closed. However the non-vanishing of the trace of the invariant tensor ϵTR, which played an important role in the previous section, places severe restrictions on a Q_S closed PCO. It turns out that any Lorentz invariant Q_S closed PCO leads to vanishing of all one loop amplitudes.

A Lorentz invariant Q_S closed PCO is defined as an operator Y that satisfies

- $Y = f_{\beta_1 \cdots \beta_{11}}(\lambda)\theta^{\beta_1} \cdots \theta^{\beta_{11}}$,
- $f_{\beta_1 \cdots \beta_{11}}(\lambda)$ has $J_{\lambda w}$ charge -11,
- $f_{\beta_1 \cdots \beta_{11}}(\lambda)$ is a Lorentz tensor,
- $Q_S Y = 0$.

The original proposal in [22] is the special case where the function f is given by[6]

$$f_{\beta_1 \cdots \beta_{11}} = \int [dC] C^1_{[\beta_1} \cdots C^{11}_{\beta_{11}]} \delta(C^1\lambda) \cdots \delta(C^{11}\lambda). \tag{5.90}$$

This satisfies the first three conditions, but although $Q_S Y \sim \lambda\delta(\lambda)$ the fourth bullet does not hold for (5.90).

Using the fact that f is a Lorentz tensor one finds,

$$\int [dB][d\lambda][dN] \lambda^{\alpha_1} \cdots \lambda^{\alpha_{11}} B^1_{m_1 n_1} \cdots B^{10}_{m_{10} n_{10}} f_{\beta_1 \cdots \beta_{11}}(\lambda) \delta(B^1 N) \cdots \delta(B^{10} N) \delta(J) =$$

$$c_1 (\epsilon TR)^{\alpha_1 \cdots \alpha_{11}}_{\beta_1 \cdots \beta_{11} m_1 n_1 \cdots m_{10} n_{10}}, \tag{5.91}$$

for some c_1. This follows from the fact that (ϵTR) is the unique Lorentz tensor with the indicated tensor structure. Now the crucial observation is that for functions f such that $Q_S Y = 0$ the integral (5.91) must be equal to zero. Indeed, using

$$0 = Q_S Y = f_{\beta_1 \cdots \beta_{11}}(\lambda) \lambda^{\beta_1} \theta^{\beta_2} \cdots \theta^{\beta_{11}}. \tag{5.92}$$

leads to

$$0 = \int [dB][d\lambda][dN] \lambda^{\alpha_2} \cdots \lambda^{\alpha_{11}} B^1_{m_1 n_1} \cdots B^{10}_{m_{10} n_{10}} \left(f_{\beta_1 \cdots \beta_{11}} \lambda^{\beta_1} \theta^{\beta_2} \cdots \theta^{\beta_{11}} \right)$$

[6] The C integral can be evaluated to give

$$f_{\beta_1 \cdots \beta_{11}} = (\epsilon T)^{\alpha\beta\gamma}_{\beta_1 \cdots \beta_{11}} \Lambda_{\alpha\beta\gamma}. \tag{5.89}$$

$$\delta(B^1N)\cdots\delta(B^{10}N)\delta(J)=c_1(\epsilon TR)^{\alpha_1\cdots\alpha_{11}}_{\alpha_1\beta_2\cdots\beta_{11}m_1n_1\cdots m_{10}n_{10}}\theta^{\beta_2}\cdots\theta^{\beta_{11}}. \tag{5.93}$$

The trace of (ϵTR) does not vanish as shown in section 5.2.3. Hence one cam conclude that

$$c_1=0. \tag{5.94}$$

To prove vanishing of all one-loop amplitudes the above result is not enough, because there are also zero mode integrals with derivatives on the delta functions and N insertions. After the nonzero mode integration is performed, an arbitrary amplitude is reduced to a sum of zero mode integrals, all of which are of the form

$$\mathcal{E}^{\alpha_1\cdots\alpha_{11}p_1q_1\cdots p_Lq_L}_{\beta_1\cdots\beta_{11}m_1n_1\cdots m_{10}n_{10}r_1s_1\cdots r_Ls_L} = \tag{5.95}$$

$$\int[dB][dN][d\lambda]\prod_{j=1}^{L}N^{p_jq_j}\prod_{i_1=1}^{L_1}B^1_{r_{i_1}s_{i_1}}\prod_{i_2=L_1+1}^{L_1+L_2}B^2_{r_{i_2}s_{i_2}}\cdots\prod_{i_{10}=L_1+\cdots+L_9+1}^{L}B^{10}_{r_{i_{10}}s_{i_{10}}}$$
$$\lambda^{\alpha_1}\cdots\lambda^{\alpha_{11}}f_{\beta_1\cdots\beta_{11}}(\lambda)B^1_{m_1n_1}\cdots B^{10}_{m_{10}n_{10}}\delta^{(L_1)}(B^1N)\cdots\delta^{(L_{10})}(B^{10}N)\delta(J),$$

where all the fields are zero modes and $L=\sum_{P=1}^{10}L_P$ and $\delta^{(m)}(x)$ denotes the m-th derivative of $\delta(x)$. All zero mode integrands have to be of the form (2.48), (2.49) for a non-vanishing answer. In order to write down the above zero mode integrand one starts from the general functions f_B, h_B from chapter 2 and uses the following four arguments.

- For each P the total number of B^P's outside the delta functions is equal to the number of derivatives on $\delta(B^PN)$ plus one. This can be inferred from the explicit form of the b ghost, (2.38), and the Taylor expansion of the delta functions. This is reflected in (5.95) because L_P appears in two places.
- For a nonzero answer the total number of N zero modes must equal the total number of derivatives on the delta functions. This gives the restriction $L=\sum L_P$.
- One might have expected derivatives on $\delta(J)$ as well, but for a non-vanishing answer there must also be enough J zero modes, so one can always reduce the amplitude to contain only $\delta(J)$.
- Compared to (2.48) the λ dependence is less general. It is possible to restrict to this class of integrands because $f_{\beta_1\cdots\beta_{11}}(\lambda)$ is a Lorentz tensor. To see this note the OPE's of N and J with f do not introduce derivatives:

$$N^{mn}(z)f_{\beta_1\cdots\beta_{11}}(\lambda(w)) \sim \sum_{i=1}^{11}(\gamma^{mn})_{\beta_i}^{\alpha}f_{\beta_1\cdots\alpha\cdots\beta_{11}}(\lambda(w))\frac{1}{z-w}, \tag{5.96}$$

$$J(z)f_{\beta_1\cdots\beta_{11}}(\lambda(w)) \sim -11f_{\beta_1\cdots\beta_{11}}(\lambda(w))\frac{1}{z-w}, \tag{5.97}$$

where the α index is in the ith position.

Note that the free indices on $\mathcal{E}$ can be either contracted among each other or with d or θ zero modes. The integral in (5.95) can be evaluated by using the definition of B integration in (2.50). Let us call the integrand of (5.95) g and write it as

$$g(\lambda, N, J, B^P) = \lambda^{\alpha_1} \cdots \lambda^{\alpha_{11}} h^{\beta_1 \cdots \beta_{11}}_{\alpha_1 \cdots \alpha_{11}}(N, J, B^P) \prod_{P=1}^{10} \delta^{(L_P)}(B^P N) f_{\beta_1 \cdots \beta_{11}}(\lambda), \quad (5.98)$$

where h is a polynomial depending on (N, J, B) as

$$(N)^L \prod_{P=1}^{10} (B^P)^{L_P+1}. \quad (5.99)$$

It also contains other fields (e.g. θ, d) but these are suppressed.

The integrations can be performed using (2.50):

$$\int [dB][d\lambda][dN] g(\lambda, N, J, B^I) \equiv \frac{\partial}{\partial \lambda^{\alpha_1}} \cdots \frac{\partial}{\partial \lambda^{\alpha_{11}}} (\epsilon T R)^{\alpha_1 \cdots \alpha_{11}}_{\beta_1 \cdots \beta_{11} m_1 n_1 \cdots m_{10} n_{10}} \quad (5.100)$$

$$\frac{\partial}{\partial B^1_{m_1 n_1}} \cdots \frac{\partial}{\partial B^{10}_{m_{10} n_{10}}} \prod_{P=1}^{10} \left(\frac{\partial}{\partial B^P_{pq}} \frac{\partial}{\partial N^{pq}}\right)^{L_P} \lambda^{\gamma_1} \cdots \lambda^{\gamma_{11}} h^{\beta_1 \cdots \beta_{11}}_{\gamma_1 \cdots \gamma_{11}}(\lambda, N, J, B^P) =$$

$$(\epsilon T R)^{\alpha_1 \cdots \alpha_{11}}_{\beta_1 \cdots \beta_{11} m_1 n_1 \cdots m_{10} n_{10}} \frac{\partial}{\partial B^1_{m_1 n_1}} \cdots \frac{\partial}{\partial B^{10}_{m_{10} n_{10}}}$$

$$\prod_{P=1}^{10} \left(\frac{\partial}{\partial B^P_{pq}} \frac{\partial}{\partial N^{pq}}\right)^{L_P} h^{\beta_1 \cdots \beta_{11}}_{\alpha_1 \cdots \alpha_{11}}(\lambda, N, J, B^P)$$

This reduces to (2.50) with $K_I = 0$ if one chooses $f_{\beta_1 \cdots \beta_{11}}(\lambda)$ as in (5.90) and uses

$$h^{\beta_1 \cdots \beta_{11}}_{\alpha_1 \cdots \alpha_{11}} = \frac{\partial}{\partial C^1_{\beta_1}} \cdots \frac{\partial}{\partial C^{11}_{\beta_{11}}} (h_B)_{\alpha_1 \cdots \alpha_{11}}. \quad (5.101)$$

Using the above definition the integral in (5.95) can be evaluated as

$$\mathcal{E}^{\alpha_1 \cdots \alpha_{11} p_1 q_1 \cdots p_L q_L}_{\beta_1 \cdots \beta_{11} m_1 n_1 \cdots m_{10} n_{10} r_1 s_1 \cdots r_L s_L} =$$

$$c_{L_1 \cdots L_{10}} \delta^{([p_1}_{r_1} \delta^{q_1]}_{s_1} \cdots \delta^{[p_L}_{r_L} \delta^{q_L])}_{s_L} (\epsilon T R)^{\alpha_1 \cdots \alpha_{11}}_{\beta_1 \cdots \beta_{11} m_1 n_1 \cdots m_{10} n_{10}}$$

$$+\text{symmetrisation in}([r_{L_{P-1}+1}, s_{L_{P-1}+1}], \ldots, [r_{L_P}, s_{L_P}], [m_P n_P]), \quad (5.102)$$

for some constant $c_{L_1 \cdots L_{10}}$. Note the round brackets denote symmetrisation in

$$[p_1 q_1], \ldots, [p_L q_L]. \quad (5.103)$$

The second line above includes ten symmetrisations, one for each P. $\mathcal{E}$ is symmetric in these indices because they all appear on B^I. (Note that by definition $L_0 = 0$).

To get some insight how to obtain (5.102) consider the case $L_1 = L = 1$. In that case the RHS of (5.95) is given by

$$(\epsilon TR)_{m_1'n_1'\cdots m_{10}'n_{10}'}\frac{\partial}{\partial B^1_{p'q'}}\frac{\partial}{\partial N^{p'q'}}\frac{\partial}{\partial B^1_{m_1'n_1'}}\cdots\frac{\partial}{\partial B^{10}_{m_{10}'n_{10}'}}N^{pq}B^1_{r_1s_1}B^1_{m_1n_1}\cdots B^{10}_{m_{10}n_{10}}, \tag{5.104}$$

where the spinor indices on (ϵTR) are suppressed. The last nine B differentiations are trivial resulting in:

$$(\epsilon TR)_{m_1'n_1'm_2n_2\cdots m_{10}n_{10}}\frac{\partial}{\partial B^1_{p'q'}}\frac{\partial}{\partial N^{p'q'}}\frac{\partial}{\partial B^1_{m_1'n_1'}}N^{pq}B^1_{r_1s_1}B^1_{m_1n_1}. \tag{5.105}$$

Let us first perform the N differentiation followed by the last two B differentiations:

$$(\epsilon TR)_{m_1'n_1'm_2n_2\cdots m_{10}n_{10}}\frac{\partial}{\partial B^1_{pq}}\frac{\partial}{\partial B^1_{m_1'n_1'}}B^1_{r_1s_1}B^1_{m_1n_1} = \tag{5.106}$$

$$(\epsilon TR)_{m_1'n_1'm_2n_2\cdots m_{10}n_{10}}\delta^{([p}_{r_1}\delta^{q]}_{s_1}\delta^{[m_1'}_{m_1}\delta^{n_1'])}_{n_1} = \delta^{[p}_{r_1}\delta^{q]}_{s_1}(\epsilon TR)_{m_1n_1\cdots m_{10}n_{10}} + (r_1s_1 \leftrightarrow m_1n_1),$$

which agrees with (5.102). The above computation clarifies the appearance of the Kronecker delta's. It is a consequence of the fact $\frac{\partial}{\partial B^1_{pq}}$ and $\frac{\partial}{\partial N^{pq}}$ appear contracted. The symmetrisations in (5.102) follow from the product rule of differentiation.

With these preliminaries it is possible to prove that if $Q_S Y = 0$ then all one-loop amplitudes vanish:

No go theorem

$$Q_S Y = 0 \Longrightarrow c_{D_1\cdots D_{10}} = 0, \tag{5.107}$$

$$c_{D_1\cdots D_{10}} = 0 \Longrightarrow \text{all one loop amplitudes vanish}, \tag{5.108}$$

Proof of (5.107). In terms of f the condition on the LHS of (5.107) reads

$$0 = Q_S Y = f_{\beta_1\cdots\beta_{11}}(\lambda)\lambda^{\beta_1}\theta^{\beta_2}\cdots\theta^{\beta_{11}}. \tag{5.109}$$

This implies

$$0 = \mathcal{E}^{\alpha_1\cdots\alpha_{11}p_1q_1\cdots p_Lq_L}_{\alpha_1\beta_2\cdots\beta_{11}m_1n_1\cdots m_{10}n_{10}r_1s_1\cdots r_{L_1}s_{L_1}} = \tag{5.110}$$

$$c_{L_1\cdots L_{10}}\delta^{([p_1}_{r_1}\delta^{q_1]}_{s_1}\cdots\delta^{[p_L}_{r_L}\delta^{q_L])}_{s_L}(\epsilon TR)^{\alpha_1\cdots\alpha_{11}}_{\alpha_1\beta_2\cdots\beta_{11}m_1n_1\cdots m_{10}n_{10}}$$

$$+\text{symmetrisation in}([r_{L_{P-1}+1}, s_{L_{P-1}+1}], \ldots, [r_{L_P}, s_{L_P}], [m_Pn_P]),$$

Since the trace of (ϵTR) does not vanish, the invariant tensor $\mathrm{Tr}\,(\epsilon TR)$ has at least one non-vanishing component. Let us denote this index choice by hats. If one chooses

$$r_is_i = \hat{m}_P\hat{n}_P, \quad i = L_{P-1}+1, \cdots, L_P, \tag{5.111}$$

$$p_iq_i = \hat{m}_P\hat{n}_P, \quad i = L_{P-1}+1, \cdots, L_P, \tag{5.112}$$

the tensor on the RHS of (5.110) is non-vanishing. Therefore

$$c_{L_1\cdots L_{10}} = 0. \tag{5.113}$$

Proof of (5.108). As explained around (5.95) all amplitudes can be written as a sum of terms, where all terms contain a $c_{L_1\cdots L_{10}}$.

This no-go theorem can be used to prove that the Y_C as defined in (2.23) are not Q_S closed. In order to see this note that computations performed in the minimal pure spinor formalism including an integration over C have led to non-vanishing answers, see e.g. [22]. In fact even agreement with the RNS formalism has been show where possible. From the no-go theorem one can conclude Y is not Q_S closed:

$$Q_S \int [dC](C^1\theta)\cdots(C^{11}\theta)\delta(C^1\lambda)\cdots\delta(C^{11}\lambda) \neq 0. \tag{5.114}$$

This implies that the individual factors, Y_C, cannot be Q_S closed either:

$$Q_S(C\theta)\delta(C\lambda) \neq 0. \tag{5.115}$$

5.4 Proof of decoupling of unphysical states

The PCO Y is not Q_S closed, hence the standard argument for decoupling of unphysical states does not apply. However that does not mean unphysical states do not decouple. One just has to use other arguments. A proof of decoupling of unphysical states in the minimal pure spinor formalism including integrals over C and B is presented in this section. Firstly the tree-level argument is reviewed in a form that generalises to the higher loops and it is shown that Q_S exact states decouple to all orders. Secondly a new symmetry of the insertions is exposed. This symmetry follows from the particular form of the picture raising operators, Z_B and it plays a crucial role in the proof. Finally this symmetry is combined with arguments based on Lorentz invariance to prove decoupling of unphysical states at every genus.

5.4.1 Tree-level amplitudes

After integrating out the nonzero modes every tree-level amplitude assumes the form

$$\mathcal{A} = \int [d\lambda][dC]d^{16}\theta\lambda^\alpha\lambda^\beta\lambda^\gamma f_{\alpha\beta\gamma}(\theta,a,k)\theta^{\beta_1}\cdots\theta^{\beta_{11}}C^1_{\beta_1}\cdots C^{11}_{\beta_{11}}\delta(C^1\lambda)\cdots\delta(C^{11}\lambda), \tag{5.116}$$

where a denotes all polarisations and k denotes all momenta. Note that the integration over the nonzero modes does not affect the factor of $Y_{C^1}\cdots Y_{C^{11}}$. This can be justified either by writing Y_C as a function of only zero modes or by inserting the factor of $(Y_C)^{11}$ at $z=\infty$ on the worldsheet. The three factors of λ originate from the three unintegrated vertex operators and the factors of θ, C and $\delta(C\lambda)$ from the eleven picture changing operators Y_C. In order to evaluate (5.116) first note that only terms with five θ's can contribute:

$$\mathcal{A} = \int [d\lambda][dC] d^{16}\theta \lambda^\alpha \lambda^\beta \lambda^\gamma f^{(5)}_{\alpha\beta\gamma\beta_{12}\cdots\beta_{16}}(a,k) \tag{5.117}$$

$$\theta^{\beta_1}\ldots\theta^{\beta_{16}} C^1_{\beta_1}\cdots C_{\beta_{11}}\delta(C^1\lambda)\cdots\delta(C^{11}\lambda).$$

The next step is showing that the integration is a projection on the scalar in $f^{(5)}$. To this end the tensor product $(\lambda)^3(\theta)^5$ is written in terms of its irreducible representations:

$$\lambda^\alpha\lambda^\beta\lambda^\gamma\theta^{\beta_{12}}\ldots\theta^{\beta_{16}} = T^{\alpha\beta\gamma\beta_{12}\cdots\beta_{16}} T_{\alpha'\beta'\gamma'\beta'_{12}\cdots\beta'_{16}} \lambda^{\alpha'}\lambda^{\beta'}\lambda^{\gamma'}\theta^{\beta'_{12}}\ldots\theta^{\beta'_{16}} + \tag{5.118}$$

$$(T_1)^{\alpha\beta\gamma\beta_{12}\cdots\beta_{16}[mn]}(T_1)_{\alpha'\beta'\gamma'\beta'_{12}\cdots\beta'_{16}[mn]}\lambda^{\alpha'}\lambda^{\beta'}\lambda^{\gamma'}\theta^{\beta'_{12}}\ldots\theta^{\beta'_{16}} +$$

$$\sum_{i\geq 2}(T_i)^{\alpha\beta\gamma\beta_{12}\cdots\beta_{16}x_i}(T_i)_{\alpha'\beta'\gamma'\beta'_{12}\cdots\beta'_{16}x_i}\lambda^{\alpha'}\lambda^{\beta'}\lambda^{\gamma'}\theta^{\beta'_{12}}\ldots\theta^{\beta'_{16}},$$

where x_i are the indices representing the representation. To obtain the above expansion one first needs to compute the tensor product $\mathrm{Gam}^3\mathbf{16}\otimes\mathrm{Asym}^5\mathbf{16}$. As discussed in section 3.4.1 this contains one scalar. One also finds there is one **45** in the tensor product, hence the second line. The sum in the last line runs over all the other irreducible representations in the tensor product, each one has an invariant tensor (T_i) associated to it. Furthermore all the (T_i)'s satisfy

$$T^{\alpha\beta\gamma\beta_{12}\cdots\beta_{16}}(T_i)_{\alpha\beta\gamma\beta_{12}\cdots\beta_{16}x_i} = 0. \tag{5.119}$$

This can be proved by contracting both sides of (5.118) with $T_{\alpha\beta\gamma\beta_{12}\cdots\beta_{16}}$. The integrations in (5.117) can be evaluated by Lorentz invariance:

$$\left(\int d^{16}\theta\theta^{\beta_1}\ldots\theta^{\beta_{16}}\right)\left(\int [d\lambda][dC]\lambda^\alpha\lambda^\beta\lambda^\gamma C^1_{\beta_1}\cdots C_{\beta_{11}}\delta(C^1\lambda)\cdots\delta(C^{11}\lambda)\right) =$$

$$\epsilon^{\beta_1\cdots\beta_{16}}(\epsilon T)^{\alpha\beta\gamma}_{\beta_1\cdots\beta_{11}} = T^{\alpha\beta\gamma\beta_{12}\cdots\beta_{16}} \tag{5.120}$$

After using (5.119) one sees all the non-scalar terms in (5.118) are annihilated by the integration. It is therefore a projection on the scalar as claimed. The final expression for the amplitude becomes

$$\mathcal{A} = T^{\alpha\beta\gamma\beta_{12}\cdots\beta_{16}} f^{(5)}_{\alpha\beta\gamma\beta_{12}\cdots\beta_{16}}(a,k). \tag{5.121}$$

Decoupling of Q_S exact states at tree level

After integrating out the nonzero modes, the amplitude containing a Q_S exact state becomes,

$$\int [d\lambda] d^{16}\theta (Q_S \Omega(\lambda, \theta, a, k)) \theta^{\beta_1} \cdots \theta^{\beta_{11}} C^1_{\beta_1} \cdots C_{\beta_{11}} \delta(C^1 \lambda) \cdots \delta(C^{11}\lambda), \tag{5.122}$$

for some Ω, where all fields are zero modes. The above integral will be shown to vanish for any Ω.

Since only the terms with five θ's and three λ's in $Q_S\Omega$ contribute, let us focus on terms in Ω with two λ's and six θ's. The upshot of the proof is that no Lorentz scalar can be constructed from two λ's and six θ's. Therefore there will be no scalar in $Q_S(\lambda)^2(\theta)^6$ and since the integration projects on the scalar the amplitude vanishes. In order to make this argument precise let us write:

$$\Omega|_{(\lambda)^2(\theta)^6} = \lambda^\alpha \lambda^\beta \theta^{\beta_1} \cdots \theta^{\beta_6} \tilde{f}_{\alpha\beta\beta_1\cdots\beta_6}(a, k) \tag{5.123}$$

for some $\tilde{f}$. The next step is writing the tensor product $(\lambda)^2(\theta)^6$ in terms of its irreducible representations:

$$\Omega|_{(\lambda)^2(\theta)^6} = \tilde{f}_{\alpha\beta\beta_1\cdots\beta_6}(a, k) \left(\sum_i (\tilde{T}_i)^{\alpha\beta\beta_1\cdots\beta_6 y_i} (\tilde{T}_i)_{\alpha'\beta'\beta'_1\cdots\beta'_6 y_i} \lambda^{\alpha'} \lambda^{\beta'} \theta^{\beta'_1} \cdots \theta^{\beta'_6} \right). \tag{5.124}$$

In the above formula it is important to note that there are no scalars in the tensor product of two pure spinors and six fermionic spinors. This is reflected by the fact that y_i represents (a positive number of) indices for every i. Now one can perform the Q_S transformation:

$$Q_S\Omega|_{(\lambda)^2(\theta)^6} = \tilde{f}_{\alpha\beta\beta_1\cdots\beta_6}(a, k) \tag{5.125}$$
$$\left(\sum_i (\tilde{T}_i)^{\alpha\beta\beta_1\cdots\beta_6 y_i} (\tilde{T}_i)_{\alpha'\beta'[\gamma'\beta'_2\cdots\beta'_6] y_i} \lambda^{\alpha'} \lambda^{\beta'} \lambda^{\gamma'} \theta^{\beta'_2} \cdots \theta^{\beta'_6} \right).$$

After invoking (5.120) one finds

$$\int [d\lambda] d^{16}\theta \left(Q_S \Omega|_{(\lambda)^2(\theta)^6} \right) \theta^{\beta_1} \cdots \theta^{\beta_{11}} C^1_{\beta_1} \cdots C_{\beta_{11}} \delta(C^1\lambda) \cdots \delta(C^{11}\lambda) = \tag{5.126}$$
$$\tilde{f}_{\alpha\beta\beta_1\cdots\beta_6}(a, k) \sum_i (\tilde{T}_i)^{\alpha\beta\beta_1\cdots\beta_6 y_i} (\tilde{T}_i)_{\alpha'\beta'[\gamma'\beta'_2\cdots\beta'_6] y_i} T^{\alpha'\beta'[\gamma'\beta'_2\cdots\beta'_6]} = 0$$

This vanishes because

$$T^{\alpha'\beta'[\gamma'\beta'_2\cdots\beta'_6]} = 0, \tag{5.127}$$

which follows from the statement that there are no scalars in $(\lambda)^2(\theta)^6$. This concludes the proof that (5.122) vanishes.

5.4.2 Higher-loop amplitudes

In order to prove decoupling of unphysical states at higher-loop amplitudes one can take similar steps to the tree-level case. This means that one first reduces the amplitude to a zero mode integral, which is effectively a projection onto a scalar and then one shows there is no scalar when one started with a Q_S exact state. In the higher-loop case an additional ingredient is needed for the second step which is a symmetry possessed by the integrand of the functional integral. This symmetry is closely related to the transformations in (4.50).

Additional symmetry

The amplitude prescription contains products of PCOs Z_B and Z_J. The main observation is that

$$Z_B Z_J = B_{mn}\lambda\gamma^{mn}d\ \delta(B_{mn}N^{mn})(\lambda d)\delta(J) \tag{5.128}$$

is invariant under

$$\delta B_{mn} = (\lambda\gamma_{[m})_\alpha f^\alpha_{n]}. \tag{5.129}$$

where $f^{n\alpha}$ are constants. This transformation acts on the $B_{mn}N^{mn}$ and $B_{mn}\lambda\gamma^{mn}d$ as,

$$\begin{aligned} \delta B_{mn}N^{mn} &= (\lambda\gamma_m)_\alpha f^\alpha_n(\lambda\gamma^{mn}w) = (\lambda\gamma^n f_n)(\lambda w), && (5.130)\\ \delta B_{mn}(\lambda\gamma^{mn}d) &= (\lambda\gamma_m)_\alpha f^\alpha_n(\lambda\gamma^{mn}d) = (\lambda\gamma^n f_n)(\lambda d). && (5.131) \end{aligned}$$

Since all these transformations contain either (λw) or (λd) and Z_J contains both $\delta(\lambda w)$ and λd:

$$\delta(Z_B Z_J) = 0. \tag{5.132}$$

Now recall that at genus g, $3g-3$ B's (one at genus one) enter via the b ghost. These B's are taken to be inert. The remaining $7g+3$ B's (9 at genus one) are taken to transform as in (5.129). Note that at one loop, the factor of $(Z_B)^9 Z_J$ is placed at a single point on the worldsheet. At two-loop order, the additional factor of $(Z_B)^7 Z_J$ is placed at a second point on the worldsheet. And at each additional loop order, one places the new factor of $(Z_B)^7 Z_J$ at a gth point on the worldsheet. With this choice, (5.129) is an invariance of the theory for $7g+3$ B's and the amplitudes must respect this symmetry.

One can understand the origin of this symmetry by going back to the first principles derivation of the amplitude prescription in chapter 4. As shown there, PCO insertions arise from gauge fixing the invariance due to pure spinor zero modes. The auxiliary fields in the gauge fixing terms have gauge invariances (cf. (4.50)). The symmetry (5.129) is a remnant of these invariances. This suggests that the amplitudes are also invariant under transformations of the $(3g-3)$ (one when $g=1$) factors of B involved in the b insertions[7], but this will not be proved or used here.

One-loop amplitudes

After integrating out all nonzero modes, as well as the d_α zero modes, every one-loop amplitude can be written as

$$\int [d\lambda][dN][dC][dB]d^{16}\theta\lambda^{\alpha_1}\cdots\lambda^{\alpha_{11}}B^1_{m_1n_1}\cdots B^{10}_{m_{10}n_{10}}f^{m_1n_1\cdots m_{10}n_{10}}_{\alpha_1\cdots\alpha_{11}}(\theta,a,k) \tag{5.133}$$

$$\theta^{\beta_1}\cdots\theta^{\beta_{11}}C^1_{\beta_1}\cdots C^{11}_{\beta_{11}}\delta(C^1\lambda)\cdots\delta(C^{11}\lambda)\delta(B^1N)\cdots\delta(B^{10}N)\delta(J),$$

where all fields are zero modes and the integrand is invariant under the B transformation (5.129). As in the tree amplitude, the integration over the nonzero modes does not affect the $(Y_C)^{11}$ factor since this factor can be written in terms of only zero modes. In this expression, eleven factors of λ originate as follows: one from the unintegrated vertex operator, one from Z_J and nine from the nine factors of Z_B. In general the zero mode integral can contain additional factors of the Lorentz currents N, higher powers of B and higher derivatives of $\delta(BN)$. These additional factors can be put into the form of (5.133) by integrating by parts using that $N^{pq}B_{mn}\partial\delta(BN)=-\delta^{[p}_m\delta^{q]}_n\delta(BN)$.

One can show that the integral in (5.133) is also a projection on a scalar. To see this first note that there is one scalar in Gam^{11}**16** $\otimes$ Asym^5**16** $\otimes$ Asym^{10}**45**. This implies one can write

$$\begin{aligned}&\lambda^{\alpha_1}\cdots\lambda^{\alpha_{11}}\theta^{\beta_{12}}\cdots\theta^{\beta_{16}}B^1_{m_1n_1}\cdots B^{10}_{m_{10}n_{10}}= \\ &(TR)^{\alpha_1\cdots\alpha_{11}\beta_{12}\cdots\beta_{16}}_{m_1n_1\cdots m_{10}n_{10}}\left((TR)(\lambda)^{11}(\theta)^5(B)^{10}\right)+ \\ &\sum_i(S_i)^{\alpha_1\cdots\alpha_{11}\beta_{12}\cdots\beta_{16}}_{m_1n_1\cdots m_{10}n_{10}x_i}\left(S_i(\lambda)^{11}(\theta)^5(B)^{10}\right)^{x_i},\end{aligned} \tag{5.134}$$

where the notation $\left((TR)(\lambda)^{11}(\theta)^5(B)^{10}\right)$ means that all indices of (TR) have been contracted with those of λ,θ and B and $\left(S_i(\lambda)^{11}(\theta)^5(B)^{10}\right)^{x_i}$ denotes an object that has x_i as its only free index and which transforms in some non-scalar representation. Similar to the tree-level case the invariant tensors S_i satisfy

$$((RT)(S_i))^{x_i}=0. \tag{5.135}$$

Note that since B is not a covariant tensor this is not the decomposition of a Lorentz invariant object into a lot of Lorentz invariant terms like (5.118). However this does

[7]Recall that $(3g-3)$ (one when $g=1$) of the Z_B factors are absorbed into the b-insertions.

not matter, the point of performing this expansion is that all the non scalar terms vanish due to the integration. The last point follows from (5.135) and

$$\int [d\lambda][dC][dB][dN]\lambda^{\alpha_1}\cdots\lambda^{\alpha_{11}}B^1_{m_1n_1}\cdots B^{10}_{m_{10}n_{10}}C^1_{\beta_1}\cdots C^{11}_{\beta_{11}} \tag{5.136}$$

$$\delta(C^1\lambda)\cdots\delta(C^{11}\lambda)\delta(B^1N)\cdots\delta(B^{10}N)\delta(J) = (\epsilon TR)^{\alpha_1\cdots\alpha_{11}}_{\beta_1\cdots\beta_{11}m_1n_1\cdots m_{10}n_{10}},$$

which is also a consequence of the fact there is only one Lorentz scalar in $\mathrm{Gam}^{11}\mathbf{16}\otimes \mathrm{Asym}^5\mathbf{16}\otimes\mathrm{Asym}^{10}\mathbf{45}$.

Decoupling of Q_S exact states

Decoupling of unphysical states will be shown by proving that if

$$\lambda^{\alpha_1}\cdots\lambda^{\alpha_{11}}B^1_{m_1n_1}\cdots B^{10}_{m_{10}n_{10}}f^{m_1n_1\cdots m_{10}n_{10}}_{\alpha_1\cdots\alpha_{11}}(\theta,a,k) \tag{5.137}$$

can be written as $Q_S\Omega$ where Ω is invariant under the B transformation then (5.133) vanishes.

Note Ω must contain ten λ's, six θ's and ten B's. There are two scalars in $\mathrm{Gam}^{10}\mathbf{16}\otimes\mathrm{Asym}^6\mathbf{16}\otimes\mathrm{Asym}^{10}\mathbf{45}$. Since $\mathrm{Gam}^{11}\mathbf{16}\otimes\mathrm{Asym}^5\mathbf{16}\otimes\mathrm{Asym}^{10}\mathbf{45}$ contains only a single scalar and Q_S maps scalars to scalars, there is a basis of invariant tensors such that one of the scalars is annihilated by Q_S and the other one, call it Ω_1, has a nonzero variation, $Q_S\Omega_1\neq 0$. This scalar is[8]

$$\Omega_1 = \left(T(\lambda)^3(\theta)^5\right)\left(R(B)^{10}(\lambda)^7(\theta)^1\right). \tag{5.138}$$

Here $(R(B)^{10}(\lambda)^7(\theta)^1)$ denotes the unique scalar obtained by contracting all indices of the objects involved. The state $Q_S\Omega_1$ is a candidate exact state that may not decouple. The scalar Ω_1 however is not invariant under the transformation (5.129) for nine of the ten B's. In fact, one can show that Ω_1 is invariant under the transformation (5.129) for only six of the ten B's. To see this, note that $\left(R(B)^{10}(\lambda)^7(\theta)^1\right)$ can be expressed as

$$(\lambda\gamma^{m_1\cdots m_5}\lambda)(\lambda\gamma^{m_6\cdots m_{10}}\lambda)(\lambda\gamma^{m_{11}\cdots m_{15}}\lambda)(\lambda\gamma^{m_{16}\cdots m_{20}}\theta) \tag{5.139}$$

contracted with the 20 vector indices of $(B)^{10}$. If both indices of B_{pq} are contracted with $m_1\cdots m_{15}$, then Ω_1 is invariant under the transformation (5.129) for that B since $(\lambda\gamma^{mn_1\cdots n_4}\lambda)(\lambda\gamma_m)_\alpha = 0$. However, if at least one index of B_{pq} is contracted with $m_{16}\cdots m_{20}$, then Ω_1 is not invariant under the transformation (5.129) for that B. Using the definition of $R^{\alpha_1\cdots\alpha_8}_{m_1\cdots m_{20}}$, one finds there are four B's whose indices are contracted with $m_{16}\cdots m_{20}$, so Ω_1 is invariant under the transformation (5.129) for six of the ten B's.

[8] Another possible candidate, $\left(T(\lambda)^2(\theta)^6\right)\left(R(B)^{10}(\lambda)^8\right)$, vanishes identically because of (5.127).

But since the gauge parameter must be invariant under (5.129) for nine of the ten B's, there is no way to generate Ω_1 as a possible gauge parameter. Thus one can conclude that if it is Q_S exact and invariant under the B transformation,

$$f^{m_1 n_1 \cdots m_{10} n_{10}}_{\alpha_1 \cdots \alpha_{11}}(\theta, a, k)\lambda^{\alpha_1} \cdots \lambda^{\alpha_{11}} B^1_{m_1 n_1} \cdots B^{10}_{m_{10} n_{10}} \tag{5.140}$$

does not contain any scalars constructed from eleven λ's, five θ's and ten B's. Since the integration projects on the (single) scalar the total zero mode integral vanishes. The precise argument is analogous to the steps in section 5.4.1.

Higher-loop amplitudes

The argument for $g > 1$ is exactly analogous. After integrating out all nonzero modes, as well as the zero modes of d_α, every $g > 1$ loop amplitude can be written as

$$\begin{aligned}&\int d^{16}\theta[d\lambda][dC]\lambda^{\alpha_1}\lambda^{\alpha_2}\lambda^{\alpha_3}\theta^{\beta_1}\cdots\theta^{\beta_{11}}C^1_{\beta_1}\cdots C^{11}_{\beta_{11}}\delta(C^1\lambda)\cdots\delta(C^{11}\lambda)\\ &\prod_{I=1}^{g}\Big([dB^I][dN^I]\lambda^{\alpha^I_4}\cdots\lambda^{\alpha^I_{11}}B^{1I}_{m^I_1 n^I_1}\cdots B^{10I}_{m^I_{10} n^I_{10}}\delta(B^{1I}N)\cdots\delta(B^{10I}N)\delta(J^I)\Big)\\ &\qquad f^{m^1_1 n^1_1 \cdots m^g_{10} n^g_{10}}_{\alpha_1\alpha_1\alpha_3\alpha^1_4\cdots\alpha^g_8}(\theta, a, k)\end{aligned} \tag{5.141}$$

where all fields are zero modes and the integrand is invariant under the B transformation (5.129). Now the factors λ originate from the $(7g + 3)$ factors of Z_B and the g factors of Z_J. Additional factors of N, B and derivatives of $\delta(BN)$ can be removed as in the one-loop case.

In this case the analogue of (5.136) is

$$\int [d\lambda][dC]\lambda^{\alpha_1}\lambda^{\alpha_2}\lambda^{\alpha_3}C^1_{\beta_1}\cdots C^{11}_{\beta_{11}}\delta(C^1\lambda)\cdots\delta(C^{11}\lambda), \tag{5.142}$$

$$\begin{aligned}&\prod_{I=1}^{g}\Big([dB^I][dN^I]\lambda^{\alpha^I_4}\cdots\lambda^{\alpha^I_{11}}B^{1I}_{m^I_1 n^I_1}\cdots B^{10I}_{m^I_{10} n^I_{10}}\delta(B^{1I}N)\cdots\delta(B^{10I}N)\delta(J^I)\Big)\\ &\qquad = (\epsilon T R^g)^{\alpha_1\alpha_2\alpha_3\alpha^1_4\cdots\alpha^g_{11}}_{\beta_1\cdots\beta_{11}m^1_1 n^1_1\cdots m^g_{10} n^g_{10}}\end{aligned}$$

where $(\epsilon T R^g)$ is the generalisation of (5.71) involving g factors of R.

There are g candidate Q_S exact states that may not decouple, which are the analogs of (5.138) and are given by

$$\Omega_J = \left(T(\lambda)^3(\theta)^5\right)\prod_{I=1}^{J-1}\left(R(B^I)^{10}(\lambda)^8\right)\ \left(R(B^J)^{10}(\lambda)^7(\theta)^1\right)\ \prod_{I=J+1}^{g}\left(R(B^I)^{10}(\lambda)^8\right) \tag{5.143}$$

where B^I denotes the B's associated with the Ith zero mode. As in the one-loop case, the term $\left(R(B^J)^{10}(\lambda)^7(\theta)^1\right)$ is at most invariant under six of the ten B^J transformations. But invariance under (5.129) requires invariance under seven of the ten B^J transformations.

This concludes the proof that unphysical states decouple to all orders in g.

5.5 Origin of the problems

Based on the BRST methods of chapter 4 one would expect that the PCOs are Q_S closed, since they originate from the gauge fixing term which is Q_S exact. However it has been proved in this chapter that the PCOs are not closed inside correlators. In order to explain this paradox let us go back to the first principles derivation of the amplitude prescription in chapter 4. Both the minimal and the non-minimal amplitude prescriptions were obtained by first coupling the pure spinor sigma model to topological gravity and then proceeding to BRST quantise this system. The BRST quantisation was applied to all gauge invariances, including the zero mode shifts of the worldsheet fields. As shown in this section the gauge fixing condition for these zero modes implicit in L_3 (cf. (4.46)) sets all the zero modes to $\lambda^\alpha = 0$. However including this point in the target space of the curved $\beta\gamma$ system that the pure spinor sector is, leads to anomalies. More precisely Nekrasov showed that the target space of curved $\beta\gamma$ systems is subject to certain conditions, which are necessary for conformal invariance of the worldsheet theory [21]. These conditions dictate that the point $\lambda^\alpha = 0$ cannot be part of the target space of the pure spinor sigma model.

Focussing on the tree level case for a moment the gauge fixing Lagrangian for the zero mode invariances is given by (after the BRST ghosts have been integrated out):

$$L_3' = \pi_\alpha\lambda^\alpha + \tilde{\pi}_\alpha\theta^\alpha. \tag{5.144}$$

5.5.1 Minimal formalism

To express the fact that π_α and $\tilde{\pi}_\alpha$ have eleven independent components they were parametrised as

$$\pi_\alpha = p_I C^I_\alpha, \qquad \tilde{\pi}_\alpha = \tilde{p}_I C^I_\alpha, \quad I = 1, \ldots, 11, \tag{5.145}$$

where C^I_α is a matrix that must have maximal rank. Thus the gauge fixing condition is given by

$$C^I_\alpha \lambda^a = 0. \tag{5.146}$$

The eleven constant spinors C^I_α are the ones that enter in the minimal pure spinor prescription. Indeed, using (5.145) one finds that the path integral contains

$$\int [dp_I][d\tilde{p}_I] \exp\left(p_I C^I_\alpha \lambda^a + \tilde{p}_I C^I_\alpha \theta^\alpha\right) = \prod_{I=1}^{11} (C^I_\alpha \theta^\alpha)\delta(C^I_\alpha \lambda^\alpha) \tag{5.147}$$

which are the eleven picture changing operators Y_C.

Implicit in (5.147) there is an analytic continuation in the field variables. A Weyl spinor in ten Euclidian dimensions cannot be real, hence λ is complex and in the minimal formulation only the holomorphic part appears. In equation (5.147) one analytically continues λ to be real and considers π_I to be purely imaginary. This can be done if the explicit expressions appearing in the amplitude computations are not singular. Typical integrals in the minimal formalism at tree level are of the form

$$\int_{-i\infty}^{i\infty} [dp] \int_{-\infty}^{\infty} [d\lambda] f(\lambda) e^{p_I C^I_\alpha \lambda^a} = \int_{-\infty}^{\infty} [d\lambda] f(\lambda)\delta(C^1\lambda)\cdots\delta(C^{11}\lambda). \tag{5.148}$$

where $f(\lambda)$ contains λ but not its complex conjugate. For this expression to be well-defined $f(\lambda)$ should not contain any $(C^I\lambda)$ poles and moreover there should not be any poles that obstruct the analytic continuation of λ to real values.

At higher loops the conjugate momentum has zero modes as well and gauge fixing this invariance leads exactly to the insertion of PCOs Z_B, Z_J, where the tensors B_{mn} enter through the gauge fixing condition, as discussed in chapter 4. In addition, one needs a composite b field satisfying (2.34). In the minimal formulation, a solution of (2.34) is given by [55]

$$b = \frac{\lambda^\alpha G^\alpha}{C_\alpha \lambda^\alpha} \tag{5.149}$$

where G^α is given in (3.139). This is however too singular to be acceptable. One can obtain a non-singular $\tilde{b}$ field by combining the b field with the PCO and solving instead (2.36). Note that this $\tilde{b}$ field now depends on the B_{mn} constant tensors but not on C_α.

5.5.2 Non-minimal formalism

The same expression (5.144) leads to the so-called regularisation factor in (5.73). This time one has to choose π_α to be a pure spinor of opposite chirality to λ^α, usually called $\bar{\lambda}_\alpha$. This indeed has eleven independent components, as required. The field $\tilde{\pi}_\alpha$, usually called r_α, automatically follows because it is the Q_S variation of π_α,

$$r_\alpha = Q_S \bar{\lambda}_\alpha. \tag{5.150}$$

This leads to the non-minimal formalism. To see this explicitly note that the factor e^{-L_3}, which is given by

$$e^{-\bar{\lambda}_\alpha \lambda^\alpha - r_\alpha \theta^\alpha}, \tag{5.151}$$

is precisely $\mathcal{N}$. The additional factors $N_{mn}\bar{N}^{mn}+\frac{1}{4}S_{mn}\lambda\gamma^{mn}d+J\bar{J}+\frac{1}{4}S\lambda d$ originate from gauge fixing the zero modes of w_α.

Note that λ is now holomorphic and $\pi_\alpha \equiv \bar{\lambda}_\alpha$ is considered as its complex conjugate variable. Typical integrals one encounters at tree level in the non-minimal formalism are therefore

$$\int [d\lambda][d\bar{\lambda}]f(\lambda)e^{-\bar{\lambda}\lambda}. \tag{5.152}$$

At higher loop order the b field enters the amplitudes. In the non-minimal formalism, equation (2.34) has a solution that depends on both λ and $\bar{\lambda}$. It is however singular as $\bar{\lambda}\lambda \to 0$ and this causes problems to certain amplitudes as explained in section 2.2. Note that the b field does not depend on how the gauge invariances due to the zero modes of w_α are treated. This is similar to the b field in (5.149) but different from $\tilde{b}$ which depends on the gauge fixing of the invariance due to zero modes of the conjugate momentum through B_{mn}.

To summarise, the minimal and non-minimal are related by field redefinitions and an analytic continuation in field space. In particular, starting from the non-minimal formalism one obtains the minimal formalism by taking $\bar{\lambda}_\alpha = C^I_\alpha \pi^I$ and analytically continuing π^I to be imaginary while at the same time analytically continuing λ to be real. There are similar redefinitions and analytic continuations in the sector related with the conjugate momentum. Furthermore, the non-minimal b field combined with part of $\mathcal{N}$ is related to $\tilde{b}$. Clearly, the two formalisms would be equivalent if the analytic continuations had not been obstructed by singularities in the amplitudes. Finally, note that the underlying gauge choice for the invariance due to pure spinor zero modes is the same: the gauge fixed action is the same, only the reality condition of the fields is different.

5.5.3 Toy example

Given the formal equivalence between the minimal and non-minimal formalisms one may wonder why the PCOs are not Q_S closed in the minimal formalism, but the corresponding object in the non-minimal formalism is Q_S closed. This issue is discussed here by analyzing a toy example that has almost all features of the actual case. Consider the following integral

$$I = \int dx dp e^{-xp}. \tag{5.153}$$

To compare with the expressions in the previous subsection p corresponds to the BRST auxiliary field and x to the pure spinor.

If one wants to evaluate the above integral, contours have to be chosen for x and p. Choosing $p = ip_1$ and $x = x_1$ with p_1, x_1 to be real, gives

$$I = i\int dx_1 dp_1 e^{ix_1p_1} = i\int_{-\infty}^{\infty} dx_1 2\pi\delta(x_1) = 2\pi i. \tag{5.154}$$

Another choice is to consider x complex and take $p = x^*$. In this case I becomes

$$I = \int dx dx^* e^{-xx^*} = 2i \int_0^\infty r dr \int_0^{2\pi} d\theta e^{-r^2} = 2\pi i. \tag{5.155}$$

This agrees nicely with the general property of contour integrals, that one is free to deform them as long as no poles are encountered. Note that (5.154) resembles a zero mode integral in the minimal formalism and (5.155) a non-minimal one.

The difference between the two prescriptions is exposed by considering the integral I with a function f in the integrand.

$$I_{\text{min}}[f] = i \int_{-\infty}^\infty dx_1 \int_{-\infty}^\infty dp_2 e^{ix_1 p_1} f(x_1) = i \int_{-\infty}^\infty dx_1 2\pi\delta(x_1) f(x_1) = 2\pi i f(0). \tag{5.156}$$

Now rotate the contour, $p = x^*$, so that the integral becomes

$$I_{\text{non-min}}[f] = \int dx dx^* e^{-|x|^2} f(x) = 2i \int_0^\infty r dr e^{-r^2} \int_0^{2\pi} d\theta f(re^{i\theta}), \tag{5.157}$$

I_{min} is the analogue of (5.148) and $I_{\text{non-min}}$ the analogue of (5.152). I_{min} and $I_{\text{non-min}}$ give exactly the same answer if $f(x)$ is non singular but (5.156) is ill defined for any choice of singular $f(x)$ whereas (5.157) may be well defined. For example, for the function

$$f(x) = \frac{1}{x}, \tag{5.158}$$

(5.156) yields ∞ but (5.157) gives 0. More precisely, (5.157) is well defined for all functions $f(z) = \sum_n c_n z^n$, with $c_n = 0$ for $n < -1$. For the $n < -1$ terms the θ integral vanishes and the r integral diverges, which makes $I_{\text{non-min}}$ ambiguous for these kind of functions.

A third representation is obtained by noticing that the θ integral can be rewritten as a contour integral

$$\int_0^{2\pi} d\theta = -i \oint_C \frac{dz}{z} \tag{5.159}$$

where $z = re^{i\theta}$ and the contour C is a circle of radius r. Thus for any meromorphic function $f(z)$ the integral over theta is independent of r and

$$I[f] = 2i \left(\int_0^\infty r dr e^{-r^2} \right) \left(-i \oint_C \frac{dz}{z} f(z) \right) = \oint_C \frac{dz}{z} f(z) \tag{5.160}$$

The expression (5.160) are well-defined for all meromorphic functions $f(z)$ whereas (5.156) and (5.157) are not.

Going back to pure spinors and working on the patch with $\lambda^+ \neq 0$ one sees that because of the factor $(\lambda^+)^{-3}$ in the measure (cf. (3.123)) the minimal formalism is expected to have a singularity unless the integrand provides a factor of $(\lambda^+)^3$, but the expressions (5.157) and (5.160) are not necessarily singular.

5.5.4 Singular gauge and possible resolution

As mentioned in the beginning of this section the gauge (5.146) leads to $\lambda^\alpha = 0$ for any choice of the constant spinors C^I_α. To see this, recall that the space of pure spinors can be covered with sixteen coordinate patches and on each patch at least one of the components of λ^α is nonzero. Let us call this component λ^+ and solve the pure spinor condition as in (3.71). Then,

$$0 = C^I_\alpha \lambda^\alpha = C^I_+\lambda^+ + C^{I,ab}\lambda_{ab} + C^I_a\lambda^a = C^I_+\lambda^+ + C^{I,ab}\lambda_{ab} + \frac{1}{8}C^I_a\epsilon^{abcde}\lambda_{bc}\lambda_{de}\frac{1}{\lambda^+} \Rightarrow$$

$$C^I_+(\lambda^+)^2 + C^{I,ab}\lambda^+\lambda_{ab} + \frac{1}{8}C^I_a\epsilon^{abcde}\lambda_{bc}\lambda_{de} = 0. \tag{5.161}$$

This system of equations however does not have a solution with $\lambda^+ \neq 0$ and the gauge is singular. To see this, first solve ten of the above equations to obtain λ_{ab} as a function of λ^+. A scaling argument implies that these functions are linear in λ^+. After plugging in the relation $\lambda_{ab} = b_{ab}\lambda^+$ in the eleventh equation, one finds that λ^+ vanishes. Thus for any choice C^I_α of maximal rank, the path integral localises at the $\lambda^\alpha = 0$ locus[9], which is the point that should be excised from the pure spinor space for the theory to be non-anomalous [21].

As discussed above, the minimal and non-minimal formalisms are related by analytic continuation in field space. In the toy example in the previous subsection the analytic continuation from the "minimal variables" x_1, p_1 to the "non-minimal variables" x, x^* sets to zero certain singular contributions (functions $f(x) \sim x^{-1}$) but the integral still localises at $x = 0$. One would thus expect that the zero mode integrals in the non-minimal formalism localise at the $\lambda^\alpha = 0$ locus, as the minimal ones do, and the problems with the $\bar\lambda\lambda$ poles one encounters for certain amplitudes at three loops and higher are a manifestation of this fact.

To avoid these problems one must find a way to gauge fix the zero mode invariances such that the zero mode integrals do not localise at $\lambda^\alpha = 0$. Let us discuss how to achieve this in the minimal formulation. First, in order to avoid the unnecessary analytic continuation to real λ one should work with the analogue of the contour representation of the delta function in (5.160) which is appropriate for holomorphic λ (and is less singular than (5.156) and (5.157)). In this language the choice of C's translates into a choice of position of poles. Secondly, one must take global issues into account. In particular, as mentioned above, the space of pure spinors can be covered with sixteen coordinates patches. In order to avoid landing in the singular gauge discussed above, one should arrange such that the expression for the path integral insertions valid in any given patch always contains at least one pole that lies in another patch. Relevant related work can be found in [43].

[9]This also shows that the choice of C in (5.2) that manifestly leads to a factor $\delta(\lambda^+)$ is not special. Any other choice of C will also contain this factor.

Chapter 6

Discussion and conclusion

After a general introduction to string theory the pure spinor formalism was presented in chapter 2. Important conclusions from that chapter include

1. The pure spinor formalism possesses more computational power than its "competitors", i.e. the RNS and Green Schwarz formalisms. This is a consequence of the explicit manifestion of both spacetime supersymmetry and Lorentz invariance within the pure spinor formalism.

2. The amplitude prescription is *ad hoc*, i.e. it has not been derived from first principles and a number of aspect of the prescription are motivated by analogy to other, older string theory formalisms.

3. Chapter 2 does not provide a proof for the decoupling of unphysical states in the minimal pure spinor formalism.

First principles derivation

Point 2 was addressed in chapter 4. The first principles derivations given in that chapter confirmed the prescriptions of chapter 2, which were advocated originally in [22] and [25, 27]. Let us summarise the approach. The pure spinor model (i.e. the Green-Schwarz-Siegel action plus the pure spinor variables) was considered as a "matter" sigma model with target space ten-dimensional superspace (with embedding coordinates X, θ) times the pure spinor space (with embedding coordinates λ). To construct a string theory this model was coupled to two-dimensional (topological) gravity and subsequently quantised by adding a gauge fixing Lagrangian to the classical action. One should contrast this approach with previous works where the aim was to find a model with local symmetry which upon gauge fixing would lead to the pure spinor model with Q_S emerging as the BRST operator and the pure spinors

λ as the corresponding ghosts. In the approach of this thesis Q_S and λ are part of the model *ab initio* and the justification for starting with this model is that the cohomology of Q_S gives the superstring spectrum. To maintain the Q_S symmetry and consistently quantise the model after coupling to $2d$ gravity, the Q_S symmetry had to be extended to act on the gravitational sector and Q_S invariance requires the existence of a (composite) field G whose Q_S variation is equal to the $2d$ stress energy tensor.

To quantise this system standard BRST techniques were followed, introduced diffeomorphism ghosts, their Q_S partners, associated auxiliary fields etc. It turns out that all variables one introduces in this process can be explicitly integrated out resulting in a prescription for the scattering amplitudes involving (as usual) a number of unintegrated and a number of integrated vertex operators and $(3g-3)$ (complex) insertions of the zero modes of G. This result holds in general for any system with a nilpotent symmetry coupled to topological gravity.

The analysis included a BRST treatment of the gauge invariances due to zero modes; the presence of a zero mode implies an invariance of the action under a shift of the field by the corresponding zero mode. To gauge fix these invariances constant ghosts and corresponding auxiliary fields were introduced. In the presence of vertex operators some of these invariances are lifted. Nevertheless, one must still gauge fix all (non-compact) bosonic invariances because their presence implies that the worldsheet action does not provide the appropriate convergence factor for the integration over them. This analysis for the bosonic zero modes of the pure spinor sigma model led (among other things) to the introduction of constant auxiliary fields needed to implement the gauge fixing conditions in the path integral. Depending on the parametrisation and the reality condition of these fields one is led either to the minimal [22] or the non-minimal [25] prescription for scattering amplitudes. In the latter case the auxiliary fields can be identified with the non-minimal variables (more precisely, the zero modes of the non-minimal variables, but since these variables are cohomologically trivial their non-zero modes do not contribute to any observable). To complete the construction one needs the explicit form of the composite "b-field" G. Although the existence of a completely satisfactory G field is guaranteed by the results of [27], the actual construction is very complicated. A possible avenue towards a simpler prescription is to look for different gauge fixing conditions for the zero modes, instead of looking for less singular representatives of $[G]$ as has been done so far.

Decoupling of unphysical states

Decoupling of unphysical, i.e. Q_S exact, states is automatic if all insertion in the amplitude prescription are Q_S closed. In the minimal formalism however the picture changing operators are not Q_S closed, in spite of the fact that their Q_S variation vanishes in a distributional sense: $Q_S Y \sim x\delta(x)$ with x that depends on λ and N. Moreover since λ parametrises a curved space it is not clear what the meaning of the delta functions in the picture changing operators is. These functions can only be properly defined in a certain coordinate patch. Indeed when trying to compute tree level amplitudes using a "naive" definition of delta functions on pure spinor space without integrating over the constant spinors C, as was done in section 5.1, one finds answers that are not Lorentz invariant. When the integral over C is included the prescription becomes manifestly Lorentz invariant. This symmetry simplifies the evaluation of the pure spinor integrals and one needs not be specific about the precise definition of the picture changing operators. After integrating over C all tree-level amplitudes, which are now automatically Lorentz invariant, vanish when one of the vertex operators is Q_S exact.

Due to the lack of a global definition for the picture changing operators, it is not possible to conclude that these operators are not Q_S closed from tree-level computations alone. This conclusion can be reached by considering one-loop amplitudes. In particular the no-go theorem shows that when the complete, i.e. containing eleven convergence factors, picture changing operator is Lorentz invariant, it cannot be Q_S closed. Else all one-loop amplitudes vanish. The fact that the PCOs are not Q_S closed does not imply that Q_S exact states do not decouple. In fact an important result of this thesis, given in section 5.4, states that Q_S exact states decouple in the minimal pure spinor formalism. This results makes use of a new symmetry of the B tensors, which has a natural place in the first principles derivation of chapter 4. Moreover the discovery of this symmetry might be a first step in making minimal loop computations as efficient as their non-minimal analogs.

In the non-minimal formalism the PCOs are replaced by the regularisation factor $\mathcal{N}$. In contrast to the PCOs, $\mathcal{N}$ is Q_S closed without subtleties. In chapter 4 it was shown that both the PCOs and the regularisation factor $\mathcal{N}$ come from a proper BRST treatment of fixing the gauge invariance generated by shifting the zero modes of the worldsheet fields. The difference between the minimal and non-minimal formalism can be understood as choosing different contours for the zero modes integrations. As became apparent in section 5.5 the choice that leads to the minimal formalism gives rise to divergent integrals for a larger class of possible insertions than the non-minimal choice. Moreover the gauge condition implicit in the current formulation of the amplitude prescriptions is singular and localises the pure spinor zero mode integrals at the $\lambda^\alpha = 0$ locus, which should be excised from the pure spinor space for the theory to be non-anomalous. The three-loop problems in the non-minimal formalism could very well be due to this singular gauge choice. To avoid these problems one should reformulate the theory in a non-singular gauge.

Appendix A

Detailed computations of I_k

This appendix contains the details of the λ integrals that appear at one loop. In particular those that play a role in computations involving a Q_S exact state. A typical integral one encounters in such an amplitude is given by

$$(I_k)_{a_1\cdots a_{2k}\beta_2\cdots\beta_{11}} = \int [d\lambda]\frac{1}{(\lambda^+)^{k-2}}\lambda^{\beta_1}\lambda_{a_1a_2}\cdots\lambda_{a_{2k-1}a_{2k}}\Lambda_{\delta_1\delta_2\delta_3}(\epsilon T)^{\delta_1\delta_2\delta_3}_{\beta_1\cdots\beta_{11}}. \quad (A.1)$$

By charge conservation one can conclude that at most two choices for $\beta_2\cdots\beta_{11}$ lead to a non-vanishing I'_k for any k. This follows from

$$0 = N(I_k)_{a_1\cdots a_{2k}\beta_2\cdots\beta_{11}} = [(k-3)\frac{5}{4}+k(-\frac{1}{4})+N(\beta_2\cdots\beta_{11})](I_k)_{a_1\cdots a_{2k}\beta_2\cdots\beta_{11}}. \quad (A.2)$$

This fixes the total charge of the β indices, which implies there are only two choices. For example for $k=3$ equation (A.2) implies only the only non-vanishing components satisfy $N(\beta_2\cdots\beta_{11}) = -\frac{1}{2}$. Thus $\beta_2\cdots\beta_{11}$ must consist of either seven **10** indices and three $\bar{\mathbf{5}}$ or a $+$, five **10**'s and four $\bar{\mathbf{5}}$'s.

In section A.1 we first compute all integrals of the form

$$(I'_k)^{\beta_1}_{a_1\cdots a_{2k}\delta_1\delta_2\delta_3} = \int [d\lambda]\frac{1}{(\lambda^+)^{k-2}}\lambda^{\beta_1}\lambda_{a_1a_2}\cdots\lambda_{a_{2k-1}a_{2k}}\Lambda_{\delta_1\delta_2\delta_3}. \quad (A.3)$$

Since I_k vanishes for $k<3$ (cf. (5.51)), we are only interested in I'_k for $k\geq 3$. By a similar argument the I'_k's are also only non-vanishing for at most two choices of $\delta_1\delta_2\delta_3$. In the last subsection half of the non-vanishing components of I_3 and all components of I_5 are computed.

A.1 Coefficients in λ integrals

For a given k at most two components of $\Lambda_{\alpha\beta\gamma}$ give non-vanishing results. One can make three choices for β_1 in I'_k, all three choices lead to an integral of the form (not necessarily for the same k):

$$(I_k'')_{a_1\cdots a_{2k}\delta_1\delta_2\delta_3} = \int [d\lambda] \frac{1}{(\lambda^+)^{k-3}} \lambda_{a_1a_2}\cdots\lambda_{a_{2k-1}a_{2k}}\Lambda_{\delta_1\delta_2\delta_3}. \tag{A.4}$$

After some algebra one finds the only non-vanishing components of the I_k'''s are:

$$(I_4'')_{a_1\cdots a_8+d_1d_2} = \frac{1}{20}\epsilon_{a_1a_2a_3a_4(d_1}\epsilon_{d_2)a_5a_6a_7a_8} + 2\text{ perms}, \tag{A.5}$$

$$(I_4'')_{a_1\cdots a_8}{}^{d_1d_2d_3d_4}{}_{d_5} = \frac{1}{5}\epsilon_{a_1a_2a_3a_4d_5}\delta^{[d_1}_{a_5}\delta^{d_2]}_{a_6}\delta^{[d_3}_{a_7}\delta^{d_4]}_{a_8} + 11\text{ perms}, \tag{A.6}$$

$$-\frac{1}{20}\epsilon_{a_1a_2a_3a_4d_5}\delta^{[d_1}_{a_5}\delta^{d_2}_{a_6}\delta^{d_3}_{a_7}\delta^{d_4]}_{a_8} + 5\text{ perms}$$

$$(I_5'')_{a_1\cdots a_{10}d_1d_2}{}^{d_3d_4} = \frac{1}{20}\epsilon_{(d_1|a_1a_2a_3a_4|}\epsilon_{d_2)a_5a_6a_7a_8}\delta^{[d_3}_{a_9}\delta^{d_4]}_{a_{10}} + 14\text{ perms}, \tag{A.7}$$

$$(I_6'')_{a_1\cdots a_{12}d_1d_2d_3} = \frac{1}{60}\epsilon_{(d_1|a_1a_2a_3a_4|}\epsilon_{d_2|a_5a_6a_7a_8|}\epsilon_{d_3)a_9a_{10}a_{11}a_{12}} + 14\text{ perms}. \tag{A.8}$$

The first step to obtain these results is finding the number of invariant tensors with the appropriate symmetries, this is one in all cases but the second. Finding the coefficients requires more work, this is done in subsection A.1.1. All these coefficients are fixed by (5.24), including the overall factor. Two corollaries are

$$(I_3')^b{}_{a_1\cdots a_6d_1d_2}{}^{d_3d_4} = (5\delta^b_{(d_1}\epsilon_{d_2)a_1a_2a_3a_4}\delta^{[d_3}_{a_5}\delta^{d_4]}_{a_6} + \delta^b_{[a_5}\delta^{[d_3}_{a_6]}\delta^{d_4]}_{(d_1}\epsilon_{d_2)a_1a_2a_3a_4}) + 2\text{ perms}, \tag{A.9}$$

$$(I_4')^b{}_{a_1\cdots a_8d_1d_2d_3} = \frac{1}{12}\delta^b_{(d_1}\epsilon_{d_2|a_1a_2a_3a_4|}\epsilon_{d_3)a_5a_6a_7a_8} + 2\text{ perms}. \tag{A.10}$$

A.1.1 Proof of equations (A.5) and (A.6)

By Lorentz invariance one can write

$$\int [d\lambda]\frac{1}{\lambda^+}\lambda_{a_1a_2}\cdots\lambda_{a_7a_8}\Lambda_{+d_1d_2} = c_3\epsilon_{a_1a_2a_3a_4(d_1}\epsilon_{d_2)a_5a_6a_7a_8} + 2\text{ perms} \tag{A.11}$$

and

$$\int [d\lambda]\frac{1}{\lambda^+}\lambda_{a_1a_2}\cdots\lambda_{a_7a_8}\Lambda^{d_1d_2d_3d_4}{}_{d_5} = c_4(\epsilon_{a_1a_2a_3a_4d_5}\delta^{[d_1}_{a_5}\delta^{d_2]}_{a_6}\delta^{[d_3}_{a_7}\delta^{d_4]}_{a_8} + 11\text{ perms}) +$$

$$c_5(\epsilon_{a_1a_2a_3a_4d_5}\delta^{[d_1}_{a_5}\delta^{d_2}_{a_6}\delta^{d_3}_{a_7}\delta^{d_4]}_{a_8} + 5\text{ perms}). \tag{A.12}$$

for some coefficients c_3, c_4, c_5. They can be determined from the defining equation of $\Lambda_{\alpha\beta\gamma}$, (5.24). After evaluating the r.h.s. of that equation for the relevant components one finds

$$\int [d\lambda]\lambda^a\lambda^b\lambda^+\Lambda_{+d_1d_2} = \delta^{(a}_{d_1}\delta^{b)}_{d_2} - \frac{2}{5}\delta^{(a}_{d_1}\delta^{b)}_{d_2} = \frac{3}{5}\delta^{(a}_{d_1}\delta^{b)}_{d_2}, \tag{A.13}$$

$$\int [d\lambda]\lambda^a\lambda^b\lambda^+\Lambda^{d_1d_2d_3d_4}{}_{d_5} = \frac{1}{5}\epsilon^{d_1d_2d_3d_4(a}\delta^{b)}_{d_5}, \tag{A.14}$$

$$\int [d\lambda]\lambda_{a_1a_2}\lambda_{a_3a_4}\lambda^a\Lambda^{d_1d_2d_3d_4}{}_{d_5} = (\delta^{d_1}_{[a_1}\delta^{d_2}_{a_2]}\delta^{d_3}_{[a_3}\delta^{d_4}_{a_4]}\delta^a_{d_5} + 1\text{ perm}) \tag{A.15}$$

$$-\frac{1}{5}\delta^{d_1}_{[a_1}\delta^{d_2}_{a_2}\delta^{d_3}_{a_3}\delta^{d_4}_{a_4]}\delta^a_{d_5} + (\frac{1}{5}\delta^a_{[a_1}\delta^{[d_1}_{a_2]}\delta^{d_2]}_{d_5}\delta^{[d_3}_{a_3}\delta^{d_4]}_{a_4} + 3\text{ perms}).$$

If one now uses equations (A.11) and (A.12) to evaluate the l.h.s. of the above integrals the values of c_3, c_4, c_5 are completely determined. In fact the integrals (A.13)-(A.15) lead to more than three equations, but they include only three independent conditions as they should. To obtain c_3 one has to write out λ^a and λ^b in (A.13) and then perform all the contractions of the two ϵ's with the r.h.s. of (A.11):

$$\frac{3}{5}\delta^{(a}_{d_1}\delta^{b)}_{d_2} = \int [d\lambda]\lambda^a\lambda^b\lambda^+\Lambda_{+d_1d_2} = 12c_3\delta^{(a}_{d_1}\delta^{b)}_{d_2} \Rightarrow c_3 = \frac{1}{20}. \tag{A.16}$$

Finding c_4 and c_5 is more involved. The l.h.s. of (A.14) can be evaluated as

$$\frac{1}{5}\epsilon^{d_1d_2d_3d_4(a}\delta^{b)}_{d_5} = \int [d\lambda]\lambda^a\lambda^b\lambda^+\Lambda^{d_1d_2d_3d_4}{}_{d_5} = (4c_4 + 12c_5)\delta^{(a}_{d_5}\epsilon^{b)d_1d_2d_3d_4}. \tag{A.17}$$

This gives the first equation for c_4, c_5. In order to completely determine them, one has to work out the l.h.s. of (A.15):

$$\frac{1}{8}\epsilon^{aa_5a_6a_7a_8}\int [d\lambda]\frac{1}{\lambda^+}\lambda_{a_1a_2}\lambda_{a_3a_4}\lambda_{a_5a_6}\lambda_{a_7a_8}\Lambda^{d_1d_2d_3d_4}{}_{d_5} = \tag{A.18}$$

$$\frac{c_4}{8}\Big((24\delta^a_{d_5}\delta^{[d_1}_{a_1}\delta^{d_2]}_{a_2}\delta^{[d_3}_{a_3}\delta^{d_4]}_{a_4} + 1\text{ perm}) + 8\epsilon^{ad_1d_2d_3d_4}\epsilon_{a_1a_2a_3a_4d_5} +$$

$$16(\delta^a_{d_5}\delta^{[d_1}_{a_1}\delta^{d_2]}_{a_2}\delta^{[d_3}_{a_3}\delta^{d_4]}_{a_4} + 1\text{ perm}) + (8\delta^a_{[a_1}\delta^{[d_3}_{a_2]}\delta^{d_4]}_{d_5}\delta^{[d_1}_{a_3}\delta^{d_2]}_{a_4} + 3\text{ perm})\Big) +$$

$$\frac{c_5}{8}\Big(24\delta^a_{d_5}\delta^{[d_1}_{a_1}\delta^{d_2}_{a_2}\delta^{d_3}_{a_3}\delta^{d_4]}_{a_4} + 24\epsilon^{ad_1d_2d_3d_4}\epsilon_{a_1a_2a_3a_4d_5} + 16\delta^a_{d_5}\delta^{[d_1}_{a_1}\delta^{d_2}_{a_2}\delta^{d_3}_{a_3}\delta^{d_4]}_{a_4} +$$

$$(8\delta^a_{[a_1}\delta^{[d_3}_{a_2]}\delta^{d_4}_{d_5}\delta^{d_1}_{a_3}\delta^{d_2]}_{a_4} + 1\text{ perm})\Big).$$

To be able to read off equations for the c's one has to rewrite the invariant tensors in terms of the ones appearing in (A.15). It turns out the space of invariant tensors with the indices and symmetries of (A.15) is four dimensional. Hence the invariant tensors in (A.18) can be written out on a basis that contains the three invariant tensors that are present in (A.15) plus a fourth one, that does not lie in the span of the first three. After using

$$\epsilon^{ad_1d_2d_3d_4}\epsilon_{a_1a_2a_3a_4d_5} = \delta^a_{d_5}\delta^{[d_1}_{a_1}\delta^{d_2}_{a_2}\delta^{d_3}_{a_3}\delta^{d_4]}_{a_4} + (\delta^a_{[a_1}\delta^{[d_1}_{a_2]}\delta^{d_2}_{d_5}\delta^{d_3}_{a_3}\delta^{d_4]}_{a_4} + 1\text{ perm}), \tag{A.19}$$

(A.18) becomes

$$(5c_4\delta^a_{d_5}\delta^{[d_1}_{a_1}\delta^{d_2]}_{a_2}\delta^{[d_3}_{a_3}\delta^{d_4]}_{a_4} + 1\text{ perm}) + (c_4\delta^a_{[a_1}\delta^{[d_3}_{a_2]}\delta^{d_4]}_{d_5}\delta^{[d_1}_{a_3}\delta^{d_2]}_{a_4} + 3\text{ perm}) + \tag{A.20}$$

$$(8c_5+c_4)\delta^a_{d_5}\delta^{[d_1}_{a_1}\delta^{d_2}_{a_2}\delta^{d_3}_{a_3}\delta^{d_4]}_{a_4}+((c_4+4c_5)\delta^a_{[a_1}\delta^{[d_3}_{a_2]}\delta^{d_4}_{d_5}\delta^{d_1}_{a_3}\delta^{d_2]}_{a_4}+1\text{ perm}).$$

Now one can read off four equations for c_4, c_5 by comparing to (A.15). Combined with the equation from (A.17) this gives:

$$5c_4=1,\quad c_4+8c_5=-\frac{1}{5},\quad c_4=\frac{1}{5},\quad c_4+4c_5=0,\quad 4c_4+12c_5=\frac{1}{5}. \tag{A.21}$$

These equations are solved by

$$c_4=\frac{1}{5},\quad c_5=-\frac{1}{20}. \tag{A.22}$$

The coefficients in equations (A.7) and (A.8) follow in the same way.

A.2 Computing the I_k's

The idea of this section is simple, use the explicit form of the gamma matrices and the λ integrals (A.5)-(A.10) to evaluate I_k. In practice this involves a lot of computation. The integrals I_0, I_1, I_2 and I_6 have already been shown to vanish in chapter 5. By the charge conservation property there is only one choice of $\beta_2\cdots\beta_{11}$ for which I_5 does not vanish. For I_3 and I_4 there are two possibilities. Let us explicitly compute I_3 for

$$\beta_2,\cdots,\beta_{11}=+,c_1,c_2,c_3,c_4,b_1b_2,\cdots,b_9b_{10}. \tag{A.23}$$

The integral I_3 consists of three terms, two for $\beta_1=b_1b_2$ and one for $\beta_1=b_1$. The first of three relevant components of ϵT is given by[1]

$$(\epsilon T)^{+d_1d_2\ b_{11}b_{12}}{}_{+}{}^{b_1b_2\cdots b_9b_{10}}{}_{c_1c_2c_3c_4}= \tag{A.24}$$

$$\frac{1}{16}8(\epsilon_{10})^{b_1\cdots b_{20}}\epsilon_{c_1\cdots c_5}\gamma^{k_1d_1}{}_{b_{13}b_{14}}\gamma^{k_2d_2}{}_{b_{15}b_{16}}\gamma_{k_3}{}^{+c_5}(\gamma_{k_1k_2}{}^{k_3})_{b_{17}b_{18}b_{19}b_{20}}=$$

$$-\frac{1}{2}8(\epsilon_{10})^{b_1\cdots b_{20}}\epsilon_{c_1c_2c_3c_4b_{17}}\delta^{d_1}_{b_{14}}\delta^{d_2}_{b_{16}}\epsilon_{b_{13}b_{15}b_{18}b_{19}b_{20}}.$$

The second relevant component is given by:

$$(\epsilon T)_{d_1d_2d_3d_4}{}^{d_5\ b_{11}b_{12}}{}_{+}{}^{b_1b_2\cdots b_9b_{10}}{}_{c_1c_2c_3c_4}= \tag{A.25}$$

$$8\frac{1}{16}2(\epsilon_{10})^{b_1\cdots b_{20}}\epsilon_{c_1\cdots c_5}\gamma_{k_1d_1d_2b_{13}b_{14}}\gamma_{k_2d_3d_4b_{15}b_{16}}\gamma^{k_3d_5}_{b_{17}b_{18}}(\gamma^{k_1k_2}{}_{k_3})_{b_{19}b_{20}}{}^{c_5}$$

$$+8\frac{1}{16}(\epsilon_{10})^{b_1\cdots b_{20}}\epsilon_{c_1\cdots c_5}\gamma^{k_1c_5}_{d_1d_2}\gamma_{k_2d_3d_4b_{13}b_{14}}\gamma^{k_3d_5}_{b_{15}b_{16}}(\gamma_{k_1}{}^{k_2}{}_{k_3})_{b_{17}b_{18}b_{19}b_{20}}+(d_1d_2\leftrightarrow d_3d_4)=$$

[1] To evaluate ϵT the following convention for $\epsilon_{\beta_1\cdots\beta_{16}}$ is used, $(\epsilon_{16})_{+a_1\cdots a_5}{}^{b_1b_2\cdots b_{19}b_{20}}=(\epsilon_5)_{a_1\cdots a_5}(\epsilon_{10})^{b_1\cdots b_{20}}$

$$8\frac{1}{4}2(\epsilon_{10})^{b_1\cdots b_{20}}\epsilon_{c_1c_2c_3c_4b_{17}}\epsilon_{b_{19}d_1d_2b_{13}b_{14}}\epsilon_{b_{20}d_3d_4b_{15}b_{16}}\delta^{d_5}_{b_{18}}+$$

$$8\frac{1}{4}(\epsilon_{10})^{b_1\cdots b_{20}}\epsilon_{c_1c_2c_3c_4b_{19}}\epsilon_{d_1d_2b_{13}b_{14}[b_{20}}\epsilon_{b_{17}]d_3d_4b_{15}b_{16}}\delta^{d_5}_{b_{18}}+$$

$$8\frac{1}{4}\frac{1}{2}(\epsilon_{10})^{b_1\cdots b_{20}}\epsilon_{c_1c_2c_3c_4[d_2}\epsilon_{d_1]b_{15}b_{19}b_{20}b_{18}}\epsilon_{b_{17}d_3d_4b_{13}b_{14}}\delta^{d_5}_{b_{16}}+(d_1d_2\leftrightarrow d_3d_4).$$

The third relevant component is given by:

$$(\epsilon T)^{d_1d_2}{}_{d_3d_4\ b+}{}^{b_1\cdots b_{10}}{}_{c_1c_2c_3c_4}= \tag{A.26}$$

$$-8\frac{1}{32}(\epsilon_{10})^{b_1\cdots b_{20}}\epsilon_{c_1c_2c_3c_4b}\gamma^{k_1d_1}_{b_{11}b_{12}}\gamma^{k_2d_2}_{b_{13}b_{14}}\gamma_{k_3d_3d_4b_{15}b_{16}}(\gamma_{k_1k_2}{}^{k_3})_{b_{17}b_{18}b_{19}b_{20}}=$$

$$8\frac{1}{4}\epsilon_{c_1c_2c_3c_4b}\epsilon_{d_3d_4b_{15}b_{16}b_{17}}\epsilon_{b_{11}b_{13}b_{18}b_{19}b_{20}}\delta^{d_1}_{b_{12}}\delta^{d_2}_{b_{14}}.$$

In the above components a factor of eight is extracted coming from the $SU(5)$ decomposition (cf. (3.22)). The powers of $\frac{1}{2}$ compensate for double counting in expressions like $x_{ab}y^{ab}$ in each line. Using the explicit form of the components of (ϵT) and the λ integrals, I_3 can be written out as

$$I_3=\frac{1}{2}3\int[d\lambda]\frac{1}{\lambda^+}\lambda_{b_{11}b_{12}}\lambda_{a_1a_2}\lambda_{a_3a_4}\lambda_{a_5a_6}\Lambda_{+d_1d_2}(\epsilon T)^{+d_1d_2\ b_{11}b_{12}}{}_{+}{}^{b_1b_2\cdots b_9b_{10}}{}_{c_1c_2c_3c_4}+$$

$$\frac{1}{8}3\int[d\lambda]\frac{1}{\lambda^+}\lambda_{b_{11}b_{12}}\lambda_{a_1a_2}\lambda_{a_3a_4}\lambda_{a_5a_6}\Lambda^{d_1d_2d_3d_4}{}_{d_5}(\epsilon T)_{d_1d_2d_3d_4}{}^{d_5\ b_{11}b_{12}}{}_{+}{}^{b_1b_2\cdots b_9b_{10}}{}_{c_1c_2c_3c_4}+$$

$$3\frac{1}{2}\int[d\lambda]\frac{1}{\lambda^+}\lambda^b\lambda_{a_1a_2}\lambda_{a_3a_4}\lambda_{a_5a_6}\Lambda_{d_1d_2}{}^{d_3d_4}(\epsilon T)^{d_1d_2}{}_{d_3d_4\ b+}{}^{b_1\cdots b_{10}}{}_{c_1c_2c_3c_4}=$$

$$\frac{3}{40}\Big(\epsilon_{a_1a_2a_3a_4(d_1}\epsilon_{d_2)a_5a_6b_{11}b_{12}}+2\text{ perms}\Big)$$

$$\Big[-\frac{1}{4}(\epsilon_{10})^{b_1\cdots b_{20}}\epsilon_{c_1c_2c_3c_4b_{17}}\delta^{d_1}_{b_{14}}\delta^{d_2}_{b_{16}}\epsilon_{b_{13}b_{15}b_{18}b_{19}b_{20}}\Big]+ \tag{A.27}$$

$$\frac{3}{40}\Big(\epsilon_{a_1a_2a_3a_4d_5}\delta^{[d_1}_{a_5}\delta^{d_2]}_{a_6}\delta^{[d_3}_{b_{11}}\delta^{d_4]}_{b_{12}}+11\text{ perms}\Big)$$

$$\Big[4(\epsilon_{10})^{b_1\cdots b_{20}}\epsilon_{c_1c_2c_3c_4b_{17}}\epsilon_{b_{19}d_1d_2b_{13}b_{14}}\epsilon_{b_{20}d_3d_4b_{15}b_{16}}\delta^{d_5}_{b_{18}}+$$

$$2(\epsilon_{10})^{b_1\cdots b_{20}}\epsilon_{c_1c_2c_3c_4b_{19}}\epsilon_{d_1d_2b_{13}b_{14}[b_{20}}\epsilon_{b_{17}]d_3d_4b_{15}b_{16}}\delta^{d_5}_{b_{18}}+$$

$$(\epsilon_{10})^{b_1\cdots b_{20}}\epsilon_{c_1c_2c_3c_4[d_2}\epsilon_{d_1]b_{15}b_{19}b_{20}b_{18}}\epsilon_{b_{17}d_3d_4b_{13}b_{14}}\delta^{d_5}_{b_{18}}+(d_1d_2\leftrightarrow d_3d_4)\Big]+$$

$$\frac{3}{2}\Big(5\delta^b_{(d_1}\epsilon_{d_2)a_1a_2a_3a_4}\delta^{[d_3}_{a_5}\delta^{d_4]}_{a_6}+\delta^b_{[a_5}\delta^{[d_3}_{a_6]}\delta^{d_4]}_{(d_1}\epsilon_{d_2)a_1a_2a_3a_4}+2\text{ perms}\Big)$$

$$\Big[2\epsilon_{c_1c_2c_3c_4b}\epsilon_{d_3d_4b_{15}b_{16}b_{17}}\epsilon_{b_{11}b_{13}b_{18}b_{19}b_{20}}\delta^{d_1}_{b_{12}}\delta^{d_2}_{b_{14}}\Big]=$$

$$-\frac{3}{5}\epsilon_{a_1a_2a_3a_4b_{14}}\epsilon_{b_{16}a_5a_6b_{11}b_{12}}(\epsilon_{10})^{b_1\cdots b_{20}}\epsilon_{c_1c_2c_3c_4b_{17}}\epsilon_{b_{13}b_{15}b_{18}b_{19}b_{20}}+2\text{ perms}+$$

$$\frac{12}{5}(\epsilon_{10})^{b_1\cdots b_{20}}\epsilon_{b_{11}b_{12}a_3a_4b_{18}}\epsilon_{c_1c_2c_3c_4b_{17}}\epsilon_{b_{19}a_1a_2b_{13}b_{14}}\epsilon_{b_{20}a_5a_6b_{15}b_{16}}+2\text{ perms}+$$

$$\frac{3}{5}(\epsilon_{10})^{b_1\cdots b_{20}}\epsilon_{b_{11}b_{12}a_3a_4b_{18}}\epsilon_{c_1c_2c_3c_4b_{19}}\epsilon_{a_1a_2b_{13}b_{14}[b_{17}}\epsilon_{b_{20}]a_5a_6b_{15}b_{16}}+2\text{ perms}+$$

$$\frac{6}{5}\epsilon_{a_1a_2a_3a_4b_{16}}(\epsilon_{10})^{b_1\cdots b_{20}}\epsilon_{c_1c_2c_3c_4b_{12}}\epsilon_{b_{11}b_{15}b_{18}b_{19}b_{20}}\epsilon_{b_{17}a_5a_6b_{13}b_{14}}+2\text{ perms}+$$

$$\frac{6}{5}\epsilon_{b_{11}b_{12}a_1a_2b_{16}}(\epsilon_{10})^{b_1\cdots b_{20}}\epsilon_{c_1c_2c_3c_4[a_4}\epsilon_{a_3]b_{15}b_{18}b_{19}b_{20}}\epsilon_{a_5a_6b_{13}b_{14}b_{17}}+2\text{ perms}+$$

$$60(\epsilon_{10})^{b_1\cdots b_{20}}\epsilon_{b_{14}a_1a_2a_3a_4}\epsilon_{c_1c_2c_3c_4b_{12}}\epsilon_{a_5a_6b_{15}b_{16}b_{17}}\epsilon_{b_{11}b_{13}b_{18}b_{19}b_{20}}+2\text{ perms}+$$

$$12\epsilon_{b_{14}a_1a_2a_3a_4}(\epsilon_{10})^{b_1\cdots b_{20}}\epsilon_{c_1c_2c_3c_4[a_5}\epsilon_{a_6]b_{12}b_{15}b_{16}b_{17}}\epsilon_{b_{11}b_{13}b_{18}b_{19}b_{20}}+2\text{ perms}=$$

$$\left(-\frac{3}{5}(1)+\frac{12}{5}(-\frac{1}{2})+\frac{3}{5}(-1)+\frac{6}{5}(-1)+\frac{6}{5}(\frac{27}{2})+60(1)+12(0)\right)$$

$$\epsilon_{a_1a_2a_3a_4b_{14}}\epsilon_{b_{16}a_5a_6b_{11}b_{12}}(\epsilon_{10})^{b_1\cdots b_{20}}\epsilon_{c_1c_2c_3c_4b_{17}}\epsilon_{b_{13}b_{15}b_{18}b_{19}b_{20}}+2\text{ perms}=$$

$$\frac{129}{2}\epsilon_{a_1a_2a_3a_4b_{14}}\epsilon_{b_{16}a_5a_6b_{11}b_{12}}(\epsilon_{10})^{b_1\cdots b_{20}}\epsilon_{c_1c_2c_3c_4b_{17}}\epsilon_{b_{13}b_{15}b_{18}b_{19}b_{20}}+2\text{ perms}.$$

Since $\text{Asym}^5\mathbf{10}\otimes\text{Sym}^3\overline{\mathbf{10}}\otimes\text{Asym}^4\overline{\mathbf{5}}$ contains one scalar all seven tensors in the penultimate step are proportional to each other. The constants of proportionality are obtained by computing components.

The integral I_5 is only non-vanishing if one chooses

$$\beta_2,\cdots,\beta_{11}=b_3b_4,\cdots,b_{11}b_{12},1,2,3,4,5. \tag{A.28}$$

This component of I_5 consists of two terms, one for $\beta_1=b_1b_2$ and one for $\beta_1=+$:

$$(I_5)_{a_1\cdots a_{10}}{}^{b_3\cdots b_{12}}{}_{12345}=\int[d\lambda]\frac{1}{(\lambda^+)^2}\lambda_{a_1a_2}\cdots\lambda_{a_9a_{10}}\Lambda_{\delta_1\delta_2\delta_3}(\epsilon T)^{\delta_1\delta_2\delta_3}{}_{+}{}^{b_3\cdots b_{12}}{}_{12345}+$$
$$\frac{1}{2}\int[d\lambda]\frac{1}{(\lambda^+)^3}\lambda_{b_1b_2}\lambda_{a_1a_2}\cdots\lambda_{a_9a_{10}}\Lambda_{\delta_1\delta_2\delta_3}(\epsilon T)^{\delta_1\delta_2\delta_3\ b_1b_2b_3\cdots b_{12}}{}_{12345}. \tag{A.29}$$

The first relevant component of ϵT is given by

$$(\epsilon T)^{d_1d_2d_3\ b_1\cdots b_{12}}{}_{12345}= \tag{A.30}$$

$$-8\frac{1}{16}2(\epsilon_{10})^{b_1\cdots b_{20}}\gamma^{ad_1}_{b_{13}b_{14}}\gamma^{bd_2}_{b_{15}b_{16}}\gamma^{cd_3}_{b_{17}b_{18}}\gamma_{abcb_{19}b_{20}}{}^{+}+$$

$$-\frac{1}{16}8(\epsilon_{10})^{b_1\cdots b_{20}}\frac{1}{2}\gamma_a{}^{(d_1}{}_{+}\gamma^{|b|d_2}_{b_{13}b_{14}}\gamma^{|c|d_3)}_{b_{15}b_{16}}\gamma^a{}_{bcb_{17}b_{18}b_{19}b_{20}}=$$

$$-(\epsilon_{10})^{b_1\cdots b_{20}}\delta^a_{[b_{13}}\delta^{d_1}_{b_{14}]}\delta^b_{[b_{15}}\delta^{d_2}_{b_{16}]}\delta^c_{[b_{17}}\delta^{d_3}_{b_{18}]}(-1)\epsilon_{abcb_{19}b_{20}}+$$

$$-\frac{1}{4}(\epsilon_{10})^{b_1\cdots b_{20}}\delta_a^{(d_1}\delta^{|b|}_{[b_{13}}\delta^{d_2}_{b_{14}]}\delta^{|c|}_{[b_{15}}\delta^{d_3)}_{b_{16}]}\delta^a_{[b_{17}}\epsilon_{b_{18}]bcb_{19}b_{20}}.$$

The second relevant component is given by

$$(\epsilon T)^{d_1d_2}{}_{d_3d_4+}{}^{b_3b_4\cdots b_{11}b_{12}}{}_{12345} = \tag{A.31}$$

$$-8\frac{1}{32}(\epsilon_{10})^{b_3\cdots b_{22}}\gamma_{ad_3d_4b_{13}b_{14}}\gamma^{bd_1}_{b_{15}b_{16}}\gamma^{cd_2}_{b_{17}b_{18}}\gamma^{a}{}_{bcb_{19}b_{20}b_{21}b_{22}} =$$

$$-\frac{1}{4}(\epsilon_{10})^{b_3\cdots b_{22}}(-1)\epsilon_{ad_3d_4b_{13}b_{14}}\delta^b_{[b_{15}}\delta^{d_1}_{b_{16}]}\delta^c_{[b_{17}}\delta^{d_2}_{b_{18}]}2\delta^a_{[b_{19}}\epsilon_{b_{20}]bcb_{21}b_{22}},$$

where again a factor of eight and powers of $\frac{1}{2}$ have been extracted. The above two components of (ϵT) can be processed further to give

$$(\epsilon T)^{d_1d_2d_3\ b_1\cdots b_{12}}{}_{12345} = -8\frac{1}{2}5(\epsilon_{10})^{b_1\cdots b_{20}}\delta^{d_1}_{b_{13}}\delta^{d_2}_{b_{15}}\delta^{d_3}_{b_{17}}\epsilon_{b_{14}b_{16}b_{18}b_{19}b_{20}} \tag{A.32}$$

and

$$(\epsilon T)^{d_1d_2}{}_{d_3d_4+}{}^{b_3b_4\cdots b_{11}b_{12}}{}_{12345} = 8\frac{1}{4}(\epsilon_{10})^{b_1\cdots b_{20}}\epsilon_{b_{17}d_3d_4b_{15}b_{16}}\epsilon_{b_{18}b_1b_{13}b_{19}b_{20}}\delta^{d_1}_{b_2}\delta^{d_2}_{b_{14}}. \tag{A.33}$$

The integral I_5 becomes

$$I_5 = \int[d\lambda]\frac{1}{(\lambda^+)^3}\lambda^{\beta_1}\lambda_{a_1a_2}\cdots\lambda_{a_9a_{10}}\Lambda_{\delta_1\delta_2\delta_3}(\epsilon T)^{\delta_1\delta_2\delta_3}{}_{\beta_1}{}^{b_3\cdots b_{12}}{}_{12345} = \tag{A.34}$$

$$\frac{1}{2}3\int[d\lambda]\frac{1}{(\lambda^+)^2}\lambda_{a_1a_2}\cdots\lambda_{a_9a_{10}}\Lambda_{d_1d_2}{}^{d_3d_4}(\epsilon T)^{d_1d_2}{}_{d_3d_4\ +}{}^{b_3\cdots b_{12}}{}_{12345}+$$

$$\frac{1}{2}\int[d\lambda]\frac{1}{(\lambda^+)^3}\lambda_{b_1b_2}\lambda_{a_1a_2}\cdots\lambda_{a_9a_{10}}\Lambda_{d_1d_2d_3}(\epsilon T)^{d_1d_2d_3\ b_1\cdots b_{12}}{}_{12345} =$$

$$\frac{3}{40}(\epsilon_{(d_1|a_1a_2a_3a_4|}\epsilon_{d_2)a_5a_6a_7a_8}\delta^{[d_3}_{a_9}\delta^{d_4]}_{a_{10}} + 14\text{ perms})(\epsilon T)^{d_1d_2}{}_{d_3d_4\ +}{}^{b_3\cdots b_{12}}{}_{12345}+$$

$$\frac{1}{120}(\epsilon_{(d_1|a_1a_2a_3a_4|}\epsilon_{d_2|a_5a_6a_7a_8|}\epsilon_{d_3)a_9a_{10}b_1b_2} + 14\text{ perms})(\epsilon T)^{d_1d_2d_3\ b_1\cdots b_{12}}{}_{12345} =$$

$$\frac{3}{20}(\epsilon_{d_1a_1a_2a_3a_4}\epsilon_{d_2a_5a_6a_7a_8}\delta^{[d_3}_{a_9}\delta^{d_4]}_{a_{10}} + 14\text{ perms})(\epsilon T)^{d_1d_2}{}_{d_3d_4\ +}{}^{b_3\cdots b_{12}}{}_{12345}+$$

$$\frac{1}{20}(\epsilon_{d_1a_1a_2a_3a_4}\epsilon_{d_2a_5a_6a_7a_8}\epsilon_{d_3a_9a_{10}b_1b_2} + 14\text{ perms})(\epsilon T)^{d_1d_2d_3\ b_1\cdots b_{12}}{}_{12345} =$$

$$\frac{3}{10}(\epsilon_{d_1a_1a_2a_3a_4}\epsilon_{d_2a_5a_6a_7a_8}\delta^{[d_3}_{a_9}\delta^{d_4]}_{a_{10}} + 14\text{ perms})$$

$$[(\epsilon_{10})^{b_1\cdots b_{20}}\epsilon_{b_{17}d_3d_4b_{15}b_{16}}\epsilon_{b_{18}b_1b_{13}b_{19}b_{20}}\delta^{d_1}_{b_2}\delta^{d_2}_{b_{14}}]$$

$$-(\epsilon_{d_1a_1a_2a_3a_4}\epsilon_{d_2a_5a_6a_7a_8}\epsilon_{d_3a_9a_{10}b_1b_2} + 14\text{ perms})$$

$$[(\epsilon_{10})^{b_1\cdots b_{20}}\delta^{d_1}_{b_{13}}\delta^{d_2}_{b_{15}}\delta^{d_3}_{b_{17}}\epsilon_{b_{14}b_{16}b_{18}b_{19}b_{20}}] =$$

$$\frac{3}{5}\epsilon_{b_2a_1a_2a_3a_4}\epsilon_{b_{14}a_5a_6a_7a_8}(\epsilon_{10})^{b_1\cdots b_{20}}\epsilon_{b_{17}a_9a_{10}b_{15}b_{16}}\epsilon_{b_{18}b_1b_{13}b_{19}b_{20}} + 14 \text{ perms}$$

$$-\epsilon_{b_{13}a_1a_2a_3a_4}\epsilon_{b_{15}a_5a_6a_7a_8}\epsilon_{b_{17}a_9a_{10}b_1b_2}(\epsilon_{10})^{b_1\cdots b_{20}}\epsilon_{b_{14}b_{16}b_{18}b_{19}b_{20}} + 14 \text{ perms} =$$

$$-\frac{2}{5}\epsilon_{b_{13}a_1a_2a_3a_4}\epsilon_{b_{15}a_5a_6a_7a_8}\epsilon_{b_{17}a_9a_{10}b_1b_2}(\epsilon_{10})^{b_1\cdots b_{20}}\epsilon_{b_{14}b_{16}b_{18}b_{19}b_{20}} + 14 \text{ perms}.$$

Bibliography

[1] M. E. Peskin and D. V. Schroeder, "An Introduction to quantum field theory,". Reading, USA: Addison-Wesley (1995) 842 p.

[2] J. Polchinski, "String theory. Vol. 1: An introduction to the bosonic string,". Cambridge, UK: Univ. Pr. (1998) 402 p.

[3] E. D'Hoker and D. H. Phong, "The Geometry of String Perturbation Theory," *Rev. Mod. Phys.* **60** (1988) 917.

[4] I. A. Batalin and G. A. Vilkovisky, "Gauge Algebra and Quantization," *Phys. Lett.* **B102** (1981) 27–31.

[5] I. A. Batalin and G. A. Vilkovisky, "Quantization of Gauge Theories with Linearly Dependent Generators," *Phys. Rev.* **D28** (1983) 2567–2582.

[6] B. Craps and K. Skenderis, "Comments on BRST quantization of strings," *JHEP* **05** (2005) 001, arXiv:hep-th/0503038.

[7] L. Brink, P. Di Vecchia, and P. S. Howe, "A Locally Supersymmetric and Reparametrization Invariant Action for the Spinning String," *Phys. Lett.* **B65** (1976) 471–474.

[8] D. Friedan, E. J. Martinec, and S. H. Shenker, "Conformal Invariance, Supersymmetry and String Theory," *Nucl. Phys.* **B271** (1986) 93.

[9] S. Deser and B. Zumino, "A Complete Action for the Spinning String," *Phys. Lett.* **B65** (1976) 369–373.

[10] J. Polchinski, "String theory. Vol. 2: Superstring theory and beyond,". Cambridge, UK: Univ. Pr. (1998) 531 p.

[11] E. D'Hoker and D. H. Phong, "Lectures on Two-Loop Superstrings," arXiv:hep-th/0211111.

[12] E. D'Hoker and D. H. Phong, "Two-Loop Superstrings I, Main Formulas," *Phys. Lett.* **B529** (2002) 241–255, arXiv:hep-th/0110247.

[13] E. D'Hoker and D. H. Phong, "Two-Loop Superstrings II, The Chiral Measure on Moduli Space," *Nucl. Phys.* **B636** (2002) 3–60, arXiv:hep-th/0110283.

[14] E. D'Hoker and D. H. Phong, "Two-Loop Superstrings III, Slice Independence and Absence of Ambiguities," *Nucl. Phys.* **B636** (2002) 61–79, arXiv:hep-th/0111016.

[15] E. D'Hoker and D. H. Phong, "Two-Loop Superstrings IV, The Cosmological Constant and Modular Forms," *Nucl. Phys.* **B639** (2002) 129–181, arXiv:hep-th/0111040.

[16] E. D'Hoker, M. Gutperle, and D. H. Phong, "Two-loop superstrings and S-duality," *Nucl. Phys.* **B722** (2005) 81–118, arXiv:hep-th/0503180.

[17] E. D'Hoker and D. H. Phong, "Two-Loop Superstrings VI: Non-Renormalization Theorems and the 4-Point Function," *Nucl. Phys.* **B715** (2005) 3–90, arXiv:hep-th/0501197.

[18] E. P. Verlinde and H. L. Verlinde, "Chiral bosonization, determinants and the string partition function," *Nucl. Phys.* **B288** (1987) 357.

[19] N. Berkovits and P. S. Howe, "Ten-dimensional supergravity constraints from the pure spinor formalism for the superstring," *Nucl. Phys.* **B635** (2002) 75–105, arXiv:hep-th/0112160.

[20] C. R. Mafra, "Superstring Scattering Amplitudes with the Pure Spinor Formalism," arXiv:0902.1552 [hep-th].

[21] N. A. Nekrasov, "Lectures on curved beta-gamma systems, pure spinors, and anomalies," arXiv:hep-th/0511008.

[22] N. Berkovits, "Multiloop amplitudes and vanishing theorems using the pure spinor formalism for the superstring," *JHEP* **09** (2004) 047, arXiv:hep-th/0406055.

[23] M. Matone, L. Mazzucato, I. Oda, D. Sorokin, and M. Tonin, "The superembedding origin of the Berkovits pure spinor covariant quantization of superstrings," *Nucl. Phys.* **B639** (2002) 182–202, arXiv:hep-th/0206104.

[24] N. Berkovits, "Cohomology in the pure spinor formalism for the superstring," *JHEP* **09** (2000) 046, arXiv:hep-th/0006003.

[25] N. Berkovits, "Pure spinor formalism as an N = 2 topological string," *JHEP* **10** (2005) 089, arXiv:hep-th/0509120.

[26] O. A. Bedoya and N. Berkovits, "GGI Lectures on the Pure Spinor Formalism of the Superstring," arXiv:0910.2254 [hep-th].

[27] N. Berkovits and N. Nekrasov, "Multiloop superstring amplitudes from non-minimal pure spinor formalism," *JHEP* **12** (2006) 029, arXiv:hep-th/0609012.

[28] P. A. Grassi and P. Vanhove, "Higher-loop amplitudes in the non-minimal pure spinor formalism," arXiv:0903.3903 [hep-th].

[29] Y. Aisaka and N. Berkovits, "Pure Spinor Vertex Operators in Siegel Gauge and Loop Amplitude Regularization," arXiv:0903.3443 [hep-th].

[30] N. Berkovits and B. C. Vallilo, "Consistency of super-Poincare covariant superstring tree amplitudes," *JHEP* **07** (2000) 015, arXiv:hep-th/0004171.

[31] C. R. Mafra, "Four-point one-loop amplitude computation in the pure spinor formalism," *JHEP* **01** (2006) 075, arXiv:hep-th/0512052.

[32] N. Berkovits and C. R. Mafra, "Equivalence of two-loop superstring amplitudes in the pure spinor and RNS formalisms," *Phys. Rev. Lett.* **96** (2006) 011602, arXiv:hep-th/0509234.

[33] N. Berkovits and C. R. Mafra, "Some superstring amplitude computations with the non- minimal pure spinor formalism," *JHEP* **11** (2006) 079, arXiv:hep-th/0607187.

[34] C. R. Mafra and C. Stahn, "The One-loop Open Superstring Massless Five-point Amplitude with the Non-Minimal Pure Spinor Formalism," arXiv:0902.1539 [hep-th].

[35] N. Berkovits, "New higher-derivative R**4 theorems," *Phys. Rev. Lett.* **98** (2007) 211601, arXiv:hep-th/0609006.

[36] N. Berkovits, M. B. Green, J. G. Russo, and P. Vanhove, "Non-renormalization conditions for four-gluon scattering in supersymmetric string and field theory," *JHEP* **11** (2009) 063, arXiv:0908.1923 [hep-th].

[37] H. Gomez, "One-loop Superstring Amplitude From Integrals on Pure Spinors Space," *JHEP* **12** (2009) 034, arXiv:0910.3405 [hep-th].

[38] H. Gomez and C. R. Mafra, "The Overall Coefficient of the Two-loop Superstring Amplitude Using Pure Spinors," *JHEP* **05** (2010) 017, arXiv:1003.0678 [hep-th].

[39] H. Georgi, "LIE ALGEBRAS IN PARTICLE PHYSICS. FROM ISOSPIN TO UNIFIED THEORIES," *Front. Phys.* **54** (1982) 1–255.

[40] A. Cohen, M. van Leeuwen, and B. Lisser, "LiE v. 2.2,". http://wwwmathlabo.univ-pointiers.fr/˜maavl/LiE/.

[41] M. Cederwall, B. E. W. Nilsson, and D. Tsimpis, "The structure of maximally supersymmetric Yang-Mills theory: constraining higher-order corrections," *JHEP* **06** (2001) 034, arXiv:hep-th/0102009.

[42] N. Berkovits, "ICTP lectures on covariant quantization of the superstring," arXiv:hep-th/0209059.

[43] N. Berkovits and S. A. Cherkis, "Pure spinors are higher-dimensional twistors," *JHEP* **12** (2004) 049, arXiv:hep-th/0409243.

[44] I. Oda and M. Tonin, "Y-formalism and b ghost in the Non-minimal Pure Spinor Formalism of Superstrings," *Nucl. Phys.* **B779** (2007) 63–100, arXiv:0704.1219 [hep-th].

[45] I. Oda and M. Tonin, "The b-field in pure spinor quantization of superstrings," arXiv:hep-th/0510223.

[46] I. Oda and M. Tonin, "Y-formalism in pure spinor quantization of superstrings," *Nucl. Phys.* **B727** (2005) 176–195, arXiv:hep-th/0505277.

[47] P. A. Grassi, G. Policastro, M. Porrati, and P. van Nieuwenhuizen, "Covariant quantization of superstrings without pure spinor constraints," *JHEP* **10** (2002) 054, arXiv:hep-th/0112162.

[48] P. A. Grassi, G. Policastro, and P. van Nieuwenhuizen, "An introduction to the covariant quantization of superstrings," *Class. Quant. Grav.* **20** (2003) S395–S410, arXiv:hep-th/0302147.

[49] P. A. Grassi, G. Policastro, and P. van Nieuwenhuizen, "The quantum superstring as a WZNW model," *Nucl. Phys.* **B676** (2004) 43–63, arXiv:hep-th/0307056.

[50] I. Oda and M. Tonin, "On the Berkovits covariant quantization of GS superstring," *Phys. Lett.* **B520** (2001) 398–404, arXiv:hep-th/0109051.

[51] Y. Aisaka and Y. Kazama, "Origin of pure spinor superstring," *JHEP* **05** (2005) 046, arXiv:hep-th/0502208.

[52] N. Berkovits and D. Z. Marchioro, "Relating the Green-Schwarz and pure spinor formalisms for the superstring," *JHEP* **01** (2005) 018, arXiv:hep-th/0412198.

[53] A. Gaona and J. A. Garcia, "BFT embedding of the Green-Schwarz superstring and the pure spinor formalism," *JHEP* **09** (2005) 083, arXiv:hep-th/0507076.

[54] R. Dijkgraaf, H. L. Verlinde, and E. P. Verlinde, "Notes on topological string theory and 2-D quantum gravity,". Based on lectures given at Spring School on Strings and Quantum Gravity, Trieste, Italy, Apr 24 - May 2, 1990 and at Cargese Workshop on Random Surfaces, Quantum Gravity and Strings, Cargese, France, May 28 - Jun 1, 1990.

[55] N. Berkovits, "Relating the RNS and pure spinor formalisms for the superstring," *JHEP* **08** (2001) 026, arXiv:hep-th/0104247.

[56] I. Oda and M. Tonin, "On the b-antighost in the pure spinor quantization of superstrings," *Phys. Lett.* **B606** (2005) 218–222, arXiv:hep-th/0409052.

[57] G. Policastro and D. Tsimpis, "R**4, purified," *Class. Quant. Grav.* **23** (2006) 4753–4780, arXiv:hep-th/0603165.

Samenvatting

Fundamenten van het pure spinor formalisme

Bij het lezen van bovenstaande titel zullen slechts enkelen een idee hebben waar dit proefschrift over gaat. Deze samenvatting is voornamelijk gericht op het complement van dit groepje experts. Beginnend bij begrippen als atomen, elementaire deeltjes en zwaartekracht, zal ik toewerken naar het eigenlijke onderwerp van dit proefschrift.

Elementaire deeltjes en krachten

Het vinden van een volledige verzameling bouwstenen van de natuur is al een eeuwenoude zoektocht. Bij de oude Grieken leefde bijvoorbeeld het idee dat alles uit vier elementen bestond: water, lucht, aarde en vuur. Deze zienswijze is nog meer dan duizend jaar leidend geweest in de wereld, maar in de laatste honderden jaren zijn de ontwikkelingen op dit gebied in een stroomversnelling geraakt. Inmiddels weten we dat alle stoffen uit moleculen bestaan, denk bijvoorbeeld aan water oftewel H_2O. Deze afkorting staat voor één water molecuul en een druppel zuiver water bestaat uit heel veel (circa 10^{23}) van deze moleculen. In figuur A.1 staat linksboven een grafische weergave van een watermolecuul. Een molecuul is echter geen elementair deeltje omdat het opgebouwd is uit bepaalde bestanddelen, de atomen. In het geval van water is dat één zuurstofatoom (O) en twee waterstofatomen (H). Ondanks het feit dat de naam atoom is afgeleid van het Griekse ἀτομος, wat ondeelbaar betekent, bestaat een atoom uit een kern en daaromheendraaiende (negatief geladen) elektronen, zoals uitgebeeld in figuur A.1. De kern is opgebouwd uit (positief geladen) protonen en (elektrisch neutrale) neutronen en deze bestaan beide uit drie quarks. De elektronen en quarks zijn voorbeelden van elementaire deeltjes. In de snaartheorie is de fundamentele aanname dat alle elementaire deeltjes, waaronder elektronen en quarks, hele kleine trillende snaartjes zijn.

Deze elementaire deeltjes kunnen met elkaar wisselwerken. Als twee elektronen bij elkaar in de buurt gebracht worden, zullen ze elkaar afstoten, omdat ze allebei een negatieve elektrische lading hebben. Dit verschijnsel wordt veroorzaakt door een elektrische kracht. Een ander voorbeeld van een kracht is de magnetische kracht,

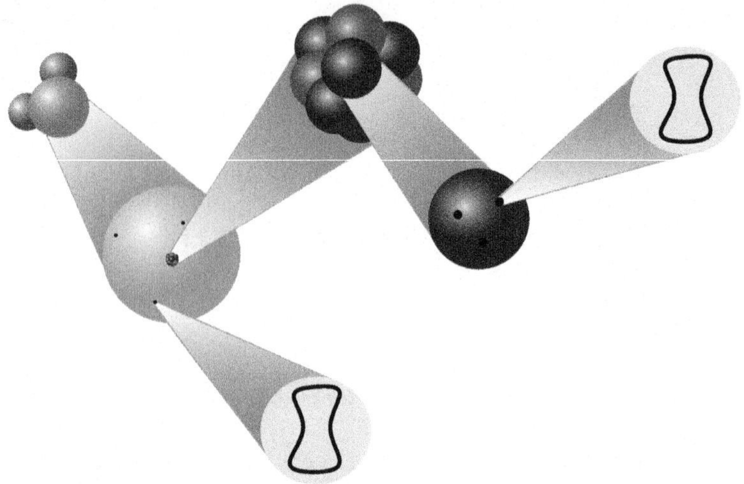

Figuur A.1: *Linksboven begint deze keten bij een molecuul, dat uit drie atomen bestaat. Elk atoom bevat een kern met daaromheen een aantal elektronen. De kern bestaat uit neutronen en protonen, die beide opgebouwd zijn uit drie quarks. In de snaartheorie worden de deeltjes aan het eind van deze keten, in dit geval elektronen en quarks, beschreven door snaren.*

denk aan twee magneten die elkaar afstoten of aantrekken, of aan de naald van een kompas. Deze twee ogenschijnlijk verschillende krachten zijn eigenlijk twee verschijningsvormen van dezelfde kracht[2], die we de *elektromagnetische* kracht noemen. Deze kracht zorgt er dus voor dat negatief en positief geladen deeltjes elkaar aantrekken, wat er onder andere toe leidt dat elektronen (negatief geladen) niet uit hun baan vliegen om de atoomkern (positief geladen). Als de elektromagnetische kracht de enige natuurkracht zou zijn, hadden atoomkernen niet kunnen bestaan, omdat deze opgebouwd zijn uit positief geladen (en neutrale) deeltjes, die elkaar elektromagnetisch afstoten. Er moet dus ook een kracht bestaan die kerndeeltjes bij elkaar houdt. Het blijken er twee te zijn: de *sterke* en de *zwakke kernkracht.* Deze drie krachten samen verklaren de wisselwerking van elementaire deeltjes, zoals gemeten in deeltjesversnellers als de LHC in Genève, tot op zeer grote precisie.

De theorie die dit soort processen beschrijft, het standaardmodel, is een zogenaamde kwantumtheorie. Het blijkt dat op deze uiterst kleine afstandschalen andere wetten gelden dan op menselijke schalen. Zo is het mogelijk dat bij een botsing van

[2]Als twee waarnemers, waarbij de één ten opzichte van de ander beweegt, naar een bepaald proces kijken, gebeurt het soms dat de één denkt een elektrische kracht te zien en de ander een magnetische. De begrippen elektrisch en magnetisch zijn dus afhankelijk van de waarnemer en daarom niet van fundamentele betekenis.

een elektron en een anti-elektron deze twee deeltjes verdwijnen en er twee andere elementaire deeltjes (bijvoorbeeld een muon en een anti-muon) verschijnen. Bovendien is het zo dat dit proces een bepaalde kans heeft, die uit het standaardmodel te bepalen is. Dit is een belangrijke eigenschap van een kwantumtheorie: er wordt niet één uitkomst met zekerheid voorspeld, maar meerdere uitkomsten elk met een bepaalde kans. Dit is geen zwaktebod van de theorie, de natuur zelf is degene die bij gelijke experimenten soms de ene uitkomst geeft en soms de andere.

Zwaartekracht

Het standaardmodel is zoals vermeld een uiterst succesvolle theorie, maar er is duidelijk meer dan elektromagnetisme en twee kernkrachten. Deze drie krachten alleen kunnen niet verklaren waarom de maan om de aarde draait, appels uit de boom vallen en er alleen naar beneden geskied kan worden. Al deze verschijnselen hebben te maken met de *zwaartekracht*. Deze kracht werkt altijd aantrekkend (in tegenstelling tot de andere drie) en speelt alleen een rol als er zeer zware objecten bij betrokken zijn. In de drie voorbeelden was dat drie keer de aarde. Dit geeft tevens een logische verklaring voor het feit dat het standaardmodel zeer nauwkeurig is ondanks dat het zwaartekracht negeert: de objecten hebben een uiterst kleine massa ten opzichte van hun elektrische lading. Deze stelling kan preciezer gemaakt worden door twee elektronen (op klassieke wijze) te beschouwen. Deze stoten elkaar elektrisch af en trekken elkaar gravitationeel aan. Voor twee elektronen op een meter afstand van elkaar zijn de krachten gegeven door:

$$\begin{aligned} F_e &= \frac{q_e^2}{4\pi\epsilon_0} = 2,3 \times 10^{-28} N, \\ F_g &= G m_e^2 = 5,5 \times 10^{-71} N, \end{aligned}$$

waar q_e de elektronlading is, e_m de elektronmassa en zowel ϵ_0 als G zijn natuurconstantes.

Dit gegeven helpt ons aan de ene kant doordat een theoretisch model voor de natuurkunde van elementaire deeltjes zonder zwaartekracht volstaat om aardse experimenten te verklaren, aan de andere kant compliceert de zwakte van de zwaartekracht de zoektocht naar een kwantumtheorie die deze kracht bevat. Simpelweg omdat we geen experimenten kunnen doen met elementaire deeltjes waar de zwaartekracht een rol speelt. Ondanks de verwaarloosbare rol van de zwaartekracht bij de controleerbare experimenten in de deeltjesversnellers zijn er redenen om een kwantumtheorie van de zwaartekracht te ontwikkelen. Er zijn namelijk situaties in de natuur waar heel veel massa zich in een zeer klein volume begeeft, bijvoorbeeld zwarte gaten of het universum vlak na de oerknal. Om deze situaties te beschrijven is een kwantumtheorie van zwaartekracht nodig. Daarnaast ligt het in de lijn der verwachting dat een goed begrip van kwantumzwaartekracht tot nieuwe inzichten zal leiden over een scala aan problemen in de natuurkunde, waarvan één van de meest in het oog springende over zogenaamde donkere materie gaat. Dit is materie waarvan het bestaan is bewezen op basis van kosmologische waarnemingen over de uitdijing van het heelal, maar het bestaat niet uit deeltjes die we kennen, d.w.z. standaardmodeldeeltjes.

Snaartheorie

Het blijkt uiterst lastig te zijn zwaartekracht in het standaardmodel in te passen. Er is dus een radicaal andere benadering nodig om een theorie van kwantumzwaartekracht op te stellen. Snaartheorie is zo een andere benadering. De fundamentele aanname van deze nieuwe theorie is dat elementaire deeltjes geen nul dimensionale objecten zijn (d.w.z. puntdeeltjes), maar hele kleine trillende snaartjes (d.w.z. één dimensionale objecten). Deze snaartjes kunnen gesloten (zoals in figuur A.1) of open zijn en het soort trilling bepaalt welk elementair deeltje het voorstelt. Op deze wijze blijkt het mogelijk te zijn de fundamentele principes van de kwantummechanica en de zwaartekracht te combineren. Snaartheorie is overigens niet alleen een theorie van kwantumzwaartekracht, maar een kwantumtheorie van alle krachten.

Symmetrieën

Zoals in alle kwantumtheorieën worden de voorspellingen in de snaartheorie ook gegeven door aan elke mogelijke uitkomst van een experiment een bepaalde kans toe te kennen. Voor het uitrekenen van deze kansen bestaan verschillende formalismes, die allemaal tot hetzelfde antwoord (moeten) leiden. Veel van de berekeningen in de snaartheorie zijn bijzonder gecompliceerd. Echter, het probleem op een slimme manier aan pakken kan leiden tot grote vereenvoudigingen. Een voorbeeld van zo een manier is gebruik maken van symmetrieën, bijvoorbeeld Lorentz symmetrie. Dit houdt in dat de theoretische voorspelling van een kans op een bepaald proces niet mag afhangen van de snelheid van de waarnemer. Dit legt (soms sterk) beperkende voorwaarden op de mogelijke uitkomsten van de berekeningen. Een andere symmetrie die een grote rol speelt in de snaartheorie is supersymmetrie.

Pure spinor formalisme

In alle snaartheorieformalismes kunnen we gebruik maken van het feit dat de antwoorden invariant zijn onder de Lorentz- en supertransformaties maar het komt vaak voor dat tussenstappen niet invariant onder de symmetrieën. Het is bijvoorbeeld mogelijk dat er meerdere termen ontstaan die niet afzonderlijk invariant zijn onder de symmetrie, alleen de som heeft deze eigenschap. In het pure spinor formalisme zijn de twee genoemde symmetrieën manifest in alle tussenstappen en dit resulteert in een aanzienlijke vereenvoudiging. Het pure spinor formalisme blijkt dan ook krachtiger te zijn dan haar voorgangers.

Dit proefschrift

In het eerste hoofdstuk van dit proefschrift wordt één van de oudere formalismes (RNS) van de supersymmetrische snaartheorie geïntroduceerd, waarbij er nadruk ligt op punten die moeilijkheden veroorzaken en op elementen die van belang zijn voor het pure spinor formalisme. Het eerste hoofdstuk eindigt met een uiteenzetting van de meest pregnante problemen van het RNS formalisme. In hoofdstuk twee geef ik een inleiding in het pure spinor formalisme en wordt het duidelijk hoe de problemen van het RNS formalisme hier vermeden worden. In dit hoofdstuk blijkt ook dat toen het pure spinor formalisme voor het eerst werd opgeschreven, het niet is afgeleid uit fundamentele principes, maar op basis van analogieën en intuïtie is bedacht. Dit heeft ertoe geleid dat bepaalde noodzakelijke eigenschappen van het pure spinor formalisme niet bewezen waren, alhoewel ze wel vermoed werden. In hoofdstuk vier presenteer ik een afleiding van het pure spinor formalisme vanuit fundamentele principes. Door middel van deze afleiding is het mogelijk te laten zien dat het pure spinor formalisme inderdaad de eerdergenoemde eigenschappen heeft. Hoofdstuk drie introduceert het wiskundige gereedschap, dat nodig is voor de afleiding in hoofdstuk vier. Het onderwerp van het laatste hoofdstuk voor de conclusie is de ontkoppeling van niet-fysische toestanden. Dit zijn toestanden die niet worden waargenomen in de natuur, maar wel worden meegenomen in de theorie om bepaalde symmetrieën te behouden in de tussenstappen. Achteraf moet er dan gecontroleerd worden dat het op geen enkele wijze mogelijk is, dat zo een niet-fysische toestand geproduceerd wordt bij een botsingsproces van fysische toestanden. Als dit inderdaad niet het geval is, spreekt men van ontkoppeling van niet-fysische toestanden. Hoofdstuk 5 bevat het bewijs van deze eigenschap voor het pure spinor formalisme.

Acknowledgments

During my years as a PhD student, which have led to this thesis, I have benefited from the interaction with a large number of people. Let me begin by expressing my sincere gratitude to my supervisor Kostas Skenderis. The many conversations we had have been invaluable. In the beginning they provided useful handles in the vast and complex world that string theory is, in the later stages they have played an indispensable role in the process of understanding and solving the problems that are now in our publications. On a more practical note I greatly appreciated that I could always walk into your office. Kostas, thank you very much for all this.

My other co-author, Nathan Berkovits, also deserves special mention. I have enjoyed the the time we have been working on our paper together and I would like to thank you for sharing your ideas with me during that period. Also I appreciated your comments on my other work. Finally my endeavour into the pure spinor formalism has been eased by my conversations with Carlos Mafra, both in person and via email.

Another aspect that comes to mind when I think back on the past five years are the numerous enlightening, entertaining, useful and not so useful conversations I have had over the years with my colleagues. So, Leo, Meindert, Sheer, Lotte, Jan, Ingmar, Xerxes, Ilies, Miranda, Adam, Marco, Balt, Jelena, Milena, Ardian, Johannes, Thomas, Paul, Borun and others thank you for creating the atmosphere that made this possible. I would also like to thank Lotty, Paula, Bianca, Jonneke, Sandra and Yocklang for administrational support.

Last but not least I thank my family for their continuous interest and support.

Lightning Source UK Ltd.
Milton Keynes UK
UKOW05f2240280916

284042UK00001B/78/P

9 789056 296414